Aristotle Today

Essays on Aristotle's
Ideal of Science

Aristotle Today

*Essays on Aristotle's
Ideal of Science*

Edited by Mohan Matthen

Published by
Academic Printing & Publishing
Edmonton, Alberta, Canada

Canadian Cataloguing in Publication Data

Main entry under title:
 Aristotle Today

 Based on papers originally presented at a conference
held in Edmonton in March 1986.
 Includes bibliographical references and indexes.
 ISBN 0-920980-17-1 (bound) − ISBN 0-920980-19-8 (pbk.)

 1. Aristotle.   I. Matthen, Mohan.
B485.A75 1987   185   C87-091374-3

© 1987 Academic Printing & Publishing
Cover design by Ottilie Sanderson
Illustration by Donna McKinnon
Printed by Art Design Printing Inc., Edmonton, Alberta, Canada

Table of Contents

Preface

There is hardly another major philosopher in whose work science and philosophy are so intimately intertwined than Aristotle. In notable contrast to even other scientifically minded philosophers — Descartes, Leibniz, Locke, Mach, and contemporary philosophers of science — Aristotle makes both areas of study influence the other. His conceptions of *form, essence, cause,* and *knowledge* are philosophical concepts, and they are responses to what Aristotle sees as problems of explanation in the sciences. Once having formulated these conceptions he is scrupulous in making his science use them and conform to the philosophical strictures he has laid down to govern their use. Other scientifically aware philosophers have not been able to achieve this sort of integration between the two sorts of endeavour. Even Mach, who perhaps came closest, drew his ideas from empiricism and applied them whole to science, with no special attention to serving, as opposed to merely dominating, the scientist. The application was knowledgeable, ingenious, and fruitful; but in the end it was dogmatic and unmotivated by concerns that were scientific in nature.

There are signs that Aristotle's sort of integration between science and philosophy is needed again today. Recent philosophical discussions of evolutionary theory, reductionism, space and time, quantum mechanics, cognition, thought, language, and neuroscience indicate that the bad old days when philosophers showed a slavish reverence for science are gone or going — not to mention the worse old days when they displayed an academically sterile ignorance of it, or even more ridiculously, an effete contempt for science and its doings. Aristotle is of relevance to those who share in the new integrative movement, not only because his example is an inspiring one that can be appreciated by Deans of Arts, grant-giving Councils, and who knows, perhaps even the House of Commons, but (need I say, more importantly?) because many of his ideas about the structure and methodology of science are still valid today.

It was to explore this aspect of Aristotle's philosophical persona, and its contemporary relevance, that a small conference, entitled 'Aristotle Today,' was arranged. The invitees were asked to provide for their own travel expenditures, and so we were confined to philosophers who live relatively close by. (Jonathan Barnes was visiting the University of Alberta for a month.) The conference was held in Edmonton on March 13-15, 1986. At first it appeared that the papers might be too diverse in their topics to admit of a unified volume of proceedings, but it turned out that this was not so. All sorts of quite unexpected dove-tailings emerged in the conference discussions and in subsequent re-writes. Many of these are explored in the Introduction.

Two brief notes about the form that this book has taken. It was intended, at first, that each paper should be followed by a record of the discussion at the conference. This turned out not to be appropriate in most cases, either because the paper was re-written in such a way as to make the discussion irrelevant or because the discussion did not bear retelling. The record has been included here where it did prove worthwhile to do so. Second, an attempt has been made to construct something more useful than the usual index locorum. An index locorum has been provided, but it has been constructed mechanically; every citation of an ancient text has been listed with no attempt to exclude merely incidental references. In addition, however, an 'Index of Aristotelian Interpretations' has been provided. This index lists (alphabetically by text and then numerically by Bekker page and line) all Aristotelian passages that receive significant attention, and says in a sentence what was made of them. Scanning this index should provide the reader with instant information about how whole Aristotelian books emerged from the conference, and of anything surprising or interesting done with passages from these books. I hope also that both the index of interpretations and the index of subjects will be of aid to the browser. Passages that are included in the index of interpretations are marked with an asterisk * in the index locorum.

Finally, thanks: to Steven DeHaven, Glen Koehn, and Martin Tweedale for help in organizing the conference, to DeHaven in particular for help with the indices, and to Tweedale for recording minutes of the discussion sessions following the papers; to the Department of Philosophy, Dean of Arts, and Conference Fund of the University of Alberta for financially supporting the conference; and to the Social Sciences and Humanities Research Council of Canada for research grants that helped support my work on the production of this book.

Mohan Matthen

Introduction: The Structure of Aristotelian Science

All the papers in this volume relate, in one way or another, to Aristotle's conception of a demonstrative science, or to his correlative conception of the structure of things in the real world. In this introduction, I shall attempt an exposition of certain aspects of Aristotle's conception, and to discuss what we can learn about Aristotelian science from the authors included herein. I shall also attempt to say something about the contemporary philosophical relevance of Aristotle's conception.

I Demonstration and Knowledge

Aristotle's *Posterior Analytics* is devoted to expounding the demonstrative format for presenting a science. The point of presenting a body of scientific truths in this format is to exhibit the explanatory and causal structure of this body of truths. Scientific facts are presented as being linked by logical deductions, and if the statement of a fact *F1* appears in a correct presentation as following from the statements of *F2* and *F3*, then *F2* and *F3* will be causally sufficient for *F1* in actual fact and *prior* to *F1* in the sense that they are causally responsible for *F1*. Thus for a student (and I mean this word to be taken in a more comprehensive sense than 'pupil') to comprehend scientific facts presented in this way is for him or her simultaneously to comprehend the causal relations that link the facts. Since *knowing* a fact is, according to Aristotle, not merely knowing *that* it is true, but also knowing *why* it is true, he can say that grasping the demonstrative structure of a science is what enables us to have knowledge. Scientific *research* consists in finding things out piecemeal, both facts and their causal antecedents. But what is thus discovered is not fully understood until it has been properly systematized. Research constitutes a major part of scientific activity, of course, and it does not fit the demonstrative ideal. It follows that the format recommended by the *Posterior Analytics* is not to be applied to scientific activity as such but to science construed as a finished system.[1]

The structure of a demonstrative science is easily summarized. First, it has a domain of investigation comprehended by a *basic term*. For example, metaphysics is the study of beings, geometry of spatial magnitudes, arithmetic of units, biology of things that reproduce and nourish themselves, and so on. In addition to the domain-specifying term there will be other basic or undefined terms that specify properties or subdomains contained by the subject-domain. For example, terms like 'true,' 'x belongs to (i.e., is predicated of) y,' and some others would be undefined in metaphysics, and the notions of parts, lengths, extremities, etc., inasmuch as they apply to spatial magnitude, are basic to geometry.

Second, there will be a set of basic propositions. These fall into three classes: axioms, definitions, and hypotheses. The *axioms* are common to all the sciences and fall under the special competence of metaphysics, or first philosophy. Examples of axioms are the principle of noncontradiction, the law of the excluded middle, and the proposition that when you take equals from equals, equals remain. Axioms are described as propositions that one must know in order to know anything at all (72a16-17).

Then there are the basic propositions that are proprietary to an individual science; these are called *theses* (or *posits*, by Barnes[2]). Theses

---

1 In his well-known article, 'Aristotle's Theory of Demonstration,' *Phronesis* **14** (1969) 123-52, Jonathan Barnes claims that the demonstrative structure of science is a model of how a science ought to be taught. In a sense, of course, he is right: the object of teaching is to impart knowledge to the pupil, and the pupil cannot gain knowledge of the body of truths that constitute a science without absorbing the demonstrative structure of those truths. But why emphasize learning from a teacher over other contexts in which knowledge is acquired? Barnes emphasizes, correctly, that the demonstrative order may not agree with the order of discovery; and it follows that the *Posterior Analytics* is not to be taken as a manual of research techniques. It also follows, given the connection between the demonstrative format and knowledge, that research construed as the discovery one by one of facts and their causes is not enough to give us knowledge. But a researcher and a systematizer can collaborate to gain knowledge that was unavailable before their joint activity, knowledge that was not taught to them.

2 In his translation and commentary (Oxford: Clarendon Press 1975): I prefer not to use 'posit,' because to me the word suggests a supposition of something's existence, and as we shall see in a moment, not all theses are this.

fall into two classes: the *definitions* do not presuppose the existence of anything, and the *hypotheses* (Barnes calls them *suppositions*) do.

Definitions establish derived terms: in Euclid's *Elements* Book I, 'A point is that which has no parts' is the first of 34 definitions. The existence of what they define has to be established separately, since existence is not part of what they assert or presuppose: Aristotle says that 'we show that they exist through the axioms and what has already been demonstrated' (76b9-11). (Presumably existence is established by construction in the mathematical sciences: see, for example, the first three propositions of Euclid's *Elements*, Book I. But it is hard to see how it is *proved* in the natural sciences: perhaps by perception, that is, by recognizing a perceived thing as answering to a definition [cf. 71a17-24]. In what sense is this a demonstration of existence?[3])

---

3 It may be that all definitions are nominal until an existence-proof has been given for the definiendum, at which point they become real definitions: see T.L. Heath, *The Thirteen Books of Euclid's Elements* (New York: Dover 1956), Vol.I, 143-6. Whether this is so or not, the point remains that it is only definitions with existential import that can be used in syllogistic demonstrations, since in the syllogistic universal statements are assumed to have existential import.

   If this is correct, a demonstration that used a definition would have to be accompanied by some sort of confirmation that the definiendum is instantiated. In the natural sciences (but not in the mathematical), a perceptual recognition of something as an *F* would have to serve as confirmation; and this recognition cannot be regarded as constituting a proper demonstration. So how can we allow that there are demonstrations in the natural sciences which use definitions?

   But there are difficulties here even for the mathematical sciences, where there are genuine existence *proofs*. The propositions that have to be proved are of the form 'There are things that answer to definiendum *X*,' and propositions of this form are not recognized by the syllogistic. This and other difficulties have led Jonathan Barnes to say that the theory of demonstration was formulated by Aristotle independently of a specific formal logic that serves to analyse the concept of logical entailment. See his 'Proof and the Syllogism,' in E. Berti, ed., *Aristotle on Science: The Posterior Analytics* (The Eighth *Symposium Aristotelicum*, Padua: Editrices Antenore [1982] 17-59). This particular claim of Barnes can be accepted independently of the claims he makes about the priority of the *Posterior* over the *Prior Analytics*.

   The same difficulties serve Michael Ferejohn as an occasion for asserting, implausibly to my mind, that there must be a non-deductive stage of demonstration, a stage in which the method used is Platonic division: see his 'Definition and the Two Stages of Aristotelian Demonstration,' *Review of Metaphysics* **36** (1982) 375-95. Ferejohn has to solve all of the difficulties Barnes mentions in his non-deductive stage; thus for example he has to deny that Aristotle would

Hypotheses state basic truths about the domain of investigation, or about sorts of things in it; they 'assume one or other part of a contradictory pair, that it is or that it is not' (72a18-20). The starting-points of a science, the basic terms and the basic propositions, are known, not by demonstration, but by the 'principle of knowledge by which we recognize the definitions' (72b24-5).[4] These starting-points are immediately true, or uncaused; and they are necessarily true. Nondemonstrative knowledge of the starting-points is knowledge not merely that they are true, but also that they are immediate and necessary.

The non-basic, or derived, truths are obtained from these starting-points by means of the syllogistic. For each non-basic truth we can construct a derivation tree as follows: we start from the proposition, and let its 'ancestors' be the premisses from which it is 'immediately' derived. Above these we place the propositions from which they are derived, and so on, until we arrive at starting-points of the science. These starting-points are known non-demonstratively, as we have just seen.[5]

The immediate ancestors of a proposition in a derivation tree will be the propositions (two, given the structure of the syllogistic) from which it is immediately derived: that is, if p and q are the immediate ancestors of r, then there must be no other proposition s that can be interpolated between the ancestors and r to yield a demonstration. The phrase 'to yield a demonstration' is important: for it may be possible to interpolate a proposition between premisses and conclusion in such a way that the premisses imply the interpolated proposition, and the latter implies the conclusion. But Aristotle would not allow that such interpolations preserve the demonstrative character of a deduction, since they might not preserve an argument's parallelism to the causal structure of the world. Thus the deductive links in a demonstration stand proxy for causal links in the world, and the immediate deductive links stand proxy for immediate causal links. In this way some-

---

allow that there could be deductive existence proofs in a demonstration. Similarly, he has not only to say that the axioms have a non-deductive use but also that they *cannot* be used deductively within a demonstration (since they are not all in syllogistic form, being multiply quantified).

4 Aristotle sometimes refers to the starting-points as definitions: he is not using the term in its narrower technical sense here.

5 See Jonathan Lear, *Aristotle and Logical Theory* (Cambridge: Cambridge University Press 1980), 17, for a picture of such a tree, and chapter 2 for some of the meta-logical applications of this representation of an Aristotelian deductive argument.

one can come to *know* a proposition (which entails, recall, knowing the causal relations that explain its obtaining) simply by grasping its derivation tree and the starting-points with which it begins.

The entire science consists of all the derivation trees consolidated into one grand structure, by pulling together the nodes corresponding to a single proposition – in mathematical terminology, it is a directed graph. What occurs above a particular proposition in this grand structure is just its derivation tree, and the top-most elements in the structure are basic propositions. Below a proposition there will be all the propositions that it, together with other propositions, demonstratively implies. The parts converging on a node will record the factors responsible for what lies at that node. The paths that leave a node will track the consequences of what lies at the node. The structure of the science, then, corresponds to the entire network of causal relations in its domain.[6]

II The Causal Underpinnings of Demonstration

The causal antecedents of a fact need not be *causes* in the sense in which we are accustomed to using that word. The word that is usually translated 'cause' is 'αἰτία,' and Aristotle's use of Greek locutions of the form '*a* is an αἰτία of *b*' is considerably wider than our use of the English '*a*

---

6 The following formal definitions encapsulate the conception of an Aristotelian science that I have just laid out:

$<A, B, C>$ is a *demonstrative syllogism* if $<A, B, C>$ is a valid syllogism with A and B as premises and C as conclusion, and were A and B to be true, C would be true *because* of A and B.

C *immediately follows from* A and B if $<A, B, C>$ is a demonstrative syllogism and there are no propositions D and E such that $<A, B, D>$ and $<D, E, C>$ are demonstrative syllogisms.

X is a *proposition appropriate to a domain D* if either (i) X is a thesis appropriate to D, or (ii) X is an axiom restricted to D, or (iii) A and B are propositions appropriate to D and X immediately follows from A and B.

$<D, P, R>$ is a *science* if D is a set of individuals, P is the non-empty set of propositions appropriate to D, and R is the relation $\{ <x, y, z>: x, y, z$ are members of P & z immediately follows from x and $y \}$.

This is an elaboration of the definition given by Jonathan Barnes in 'Proof and the Syllogism.' The definition would have to be modified if Aristotle's formal syllogistic were abandoned as *the* logic of demonstration.

causes b,' or 'a is a cause of b.' As Michael Frede has argued, this is mainly because we require causes to be active: the cause of X must, we think, have *done* something, or have been a transient occurrence, that resulted in X.[7] Aristotle uses 'αἰτία' for things like the matter out of which something is made, and the end towards which the thing was directed. But the glass out of which my paper-weight is made did not do anything that resulted in the paper-weight coming into existence, nor did its paper-securing function. So these uses of 'αἰτία' do not correspond to our notion of cause. This is why I have used circumlocutions like 'causal condition' or 'causal antecedent' to translate 'αἰτία.' By 'causal condition,' I just mean 'condition given the causal relations that in fact obtain.'[8] I take it that the glass in my paper-weight is a condition of *some* sort for the existence of the paper-weight, given the causal relations that in fact obtain. And so is the function that this object has.[9]

I have emphasized the parallelism of causal and deductive structure in Aristotle's conception of a science. But which structure is prior? Does Aristotle define αἰτία in terms of demonstration, or vice versa? It will be apparent that I have been taking the notion of cause to be prior, inasmuch as I have supposed that the conditions that turn a mere deduction into a demonstration are conditions that involve the notion of cause. But the question is moot. For example, it is not at all uncommon to find, in the recent literature, attempts to define the Aristotelian conception of necessity in terms of deductive relations, even when it is found in apparently scientific, as opposed to logical, works: something like this — 'A necessitates B if there is a demonstration of B from A.' The implication, conscious or not, is that the relations of necessita-

---

7 See Michael Frede, 'The Original Notion of Cause,' in Malcolm Schofield *et al.*, eds., *Doubt and Dogmatism* (Oxford: Clarendon Press 1980) 217-49.

8 It is a peculiar fact about philosophical English that the the adjective 'causal' is appropriate to many more contexts than the transitive verb 'cause,' or the noun. Thus the term 'causal relation' can be construed as comprehending all sorts of relations that depend on how the world is causally organized: it is not restricted to the relationship of cause to effect.

9 Some have urged that 'αἰτίαι' be translated 'reason' or 'explanation' — see for example, Gregory Vlastos in 'Reasons and Causes in the *Phaedo*,' *Philosophical Review* **78** (1969) 291-325; J.M.E. Moravcsik, 'Aristotle on Adequate Explanations,' *Synthese* **28** (1974) 3-17; and Jonathan Barnes in his commentary on the *Posterior Analytics*, 96. But it seems to me that this suggests that αἰτίαι are not real entities.

tion found to hold amongst the objects of a science are inherited from the deductive and epistemological relations amongst the propositions in a science rather than the other way around. Since what has been said about relations of necessitation is easily extendible to relations of 'aitiation,' it is desirable that I should say something about why one ought to interpret Aristotle in the way in which I do.

Interpreted as depending on deductive relations amongst statements, and upon epistemological and universality constraints on the starting-points of demonstration, Aristotle's conception of demonstration seems to share a great deal with Hempel's 'deductive-nomological' theory of scientific explanation.[10] According to Hempel, a statement is explained if it is logically deducible from a set that includes some 'law-like' statements, perhaps together with some other statements. But I think it doubtful that Aristotle's constraints on starting-points can in fact be captured solely in terms of universality and epistemology; it is thoroughly infected by *causal* notions. And this means that the concept of a demonstration derives from *aitiational* constraints, and cannot be used to explicate αἰτίαι.

I argue in my article below (**Mohan Matthen**, 152-5) that Aristotelian causal links are usually open to being thwarted in an explainable way: for example, truths that we might find presented by biology as holding universally may in fact fail to hold in individual cases because of the failure of form to master recalcitrant matter. Insufficient heat will thwart the proper reproduction of animal form, for example. These departures from the norm are not merely failures of universality; they are explainable in terms of regularities concerning matter (insufficient heat will *always* have this effect). But the explanation of why form failed to master matter could not be introduced into a demonstration in the science of biology proper since to do so would involve mixing together the principles of two different sciences: biology, and the science of the elements to which the study of heat belongs. In other words, the non-universality of certain biological principles is explainable by reference to universal regularities, but not within biology.

How then are we to characterize the principles of biology? From within biology all that we can say is that they are true 'for the most part': we cannot specify or explain the exceptions. I suggest that from a meta-scientific point of view, an appropriate characterization of the

---

10 See the essays in part IV of C.G. Hempel, *Aspects of Scientific Explanation* (New York: Free Press 1965).

status of the starting-points is that they operate universally and necessarily *where not interfered with* by causes that are 'accidental' to the domain under study — that is, by causes that belong appropriately to some other domain. This stipulation finds an echo in **Francis Sparshott's** article 'Aristotle's World and Mine': he says that Aristotelian science is not concerned with accidental departures from eternally repeated causal patterns and sequences (27-30). It is this sort of specification that I think cannot be captured in terms of regularities alone, and without references to causes. But it is precisely this sort of specification that is needed before one can apply the concept of demonstration to understanding αἰτίαι.

The primary textual objection to those who would put deductive relations before causal relations is that Aristotle distinguishes demonstrations from syllogisms: 'By demonstration I mean a syllogism that produces scientific knowledge ... [It] must proceed from premises that are true, primary, immediate, better-known than, prior to, and which are αἰτίαι of the conclusion. Syllogism will be possible without these conditions, but not demonstration' (71b17-24). These are the constraints on starting-points of demonstration: it seems natural, even disregarding the arguments of the last two paragraphs, to interpret these constraints as reflecting some features of the causal structure of the real world. If this is right, Aristotle sounds less like Hempel, and more like those of Hempel's recent critics who have demanded that an explanation do more than subsume a fact under a regularity: it must, they say, adduce causes of what is to be explained. The Aristotelian demand that the premises of demonstration be primary, immediate, prior to and αἰτίαι of the conclusion seems to prefigure these critics of Hempel.

Even more striking, Aristotle's constraints on demonstration seem to find an echo in the positive account of explanation offered by one of the critics, Wesley Salmon.[11] For example, Salmon requires that there be certain ultimate terms in a complete scientific explanation, and that these terms be 'homogeneous' with respect to the explanandum, that is, there should no division of these terms into sub-classes relevant to the explanandum (36-7). For example, suppose that we use the fact that Jones smokes to explain the fact that he got cancer. If Jones also drinks, and people who smoke *and* drink run a higher risk of getting cancer than those who smoke and do not drink, the explanation will be incomplete.

---

11 See Wesley C. Salmon, *Scientific Explanation and the Causal Structure of the World* (Princeton, NJ: Princeton University Press 1984).

Now suppose that smoking tends to cause condition C, and C tends to cause cancer. People who smoke and have C will run a higher risk than those who merely smoke and, for whatever reason, do not have C. So it will be explanatorily incomplete to cite smoking as a cause of cancer without citing C. This amounts to requiring that regardless of the strength of the correlation between smoking and cancer, intermediate *causes* must be cited, a requirement that cannot be glossed away in terms of frequentist specifications like universality or high probability. It would be foolish to carry the comparison too far, but in the light of these conditions it is reasonable to say that what Aristotle is after when he requires that demonstrations appeal to immediate and prior conditions is not so very different from this constraint of Salmon's.

As we have seen, proponents of the priority of demonstration want to assert that all of Aristotle's constraints can be understood epistemologically. They are constraints that depend on how we grasp the statements, they say, not constraints that depend on the facts that correspond to them 'out there.' For example, priority is cognitive priority, immediacy is direct knowability, and so on. This interpretation is given some credence by the fact that Aristotle says that the premisses of a demonstration must be 'better-known' than the conclusion. What would this constraint be doing in a list of ontological conditions?

The trouble for the epistemological approach is that Aristotle seems to regard inferential relations themselves to be the subjects of causal explication. Thus in justifying the 'better-known' constraint that we have just mentioned, he argues simply that since our knowledge of the premisses is the cause of our knowledge of the conclusion, the premisses must be better-known than the conclusion. After all, he says, when the F in x is caused by (the F in) y, then y must be more F than x: so if the knowledge attaching to the conclusion is caused by the knowledge attaching to the premisses, then the premisses must be better-known than the conclusion (72a25-b3). It is to be gathered from arguments of this sort[12] that Aristotle supposes that knowledge is the *causal* product of certain mental processes. The constraints placed on the starting-points are constraints relevant to the operation of the causal processes that produce knowledge, rather than to epistemological requirements relating to justification. That is, Aristotle does not seem to be trying to *justify* one's confidence in (or knowledge of) a conclusion

---

12 The argument supporting the claim that the principle of non-contradiction is the 'best-known' is of the same sort: see **Alan Code's** article 'Metaphysics and Logic,' especially part II, section 1.

by appealing to the greater confidence that one attaches to the premisses from which one derives that conclusion. Rather, he is claiming that the degree to which one knows a proposition is a result of a transferrence of knowledge from premisses to conclusion, much as the warmth of the coffee I drink was transferred to it from the stove with which it was in contact. And just as the coffee is not as warm as the stove that heats it, the conclusion is not as well-known as the premisses that make it known.

The priority of the causal structure is further substantiated by the Aristotelian distinction between something's being better-known 'to us,' and better-known 'by nature' (eg. at 71b33-72a5). Something is better-known to us if it is easier for us to grasp it, for example because it is 'closer to perception.' Something is better-known by nature if we must know it before what is less well-known by nature, for example, the genus and differentiae are better-known than the species (*Topics* 141b29-34). The starting-points of a science are better-known 'by nature,' but not necessarily 'to us.'

Now, what makes something better-known by nature? Why should it be the case, for example, that we know the differentiae of dogs before we know dogs? Presumably this is a constraint on bestowing the honorific term 'knowledge' on a cognitive state: for it is clear that we can be familiar with dogs before we are familiar with their differentiae. Such a constraint cannot come from the psychology of knowing because any constraint that did would seem to fall under the rubric of familiarity 'to us': dogs are more faimiliar to us than their differentiae are, because they come first in the order in which knowledge is acquired. When we are acquiring knowledge about dogs, then, we might very well 'justify' some proposal about what the differentiae of dogs are by appeal to our perceptions of dogs. Why is this not allowed when we present a science demonstratively, that is, in the order of what is better-known by nature? It seems natural to answer: 'Because a demonstratively presented science orders facts according to causal roles, and only in virtue of this can knowledge be passed to a conclusion through the premisses.' I conclude therefore that the constraints on starting-points derives from the nature of the known rather than from the nature of our knowing. At any rate it is up to those who hold the other view to give us a clear account of what it means for something to be better known 'by nature.'

III First Philosophy

Aristotle's position is, as I have presented it, that we format a science in a certain way because it parallels the causal structure of the world. But that is not all of it. For, as we have seen, he holds as well that knowledge is of what is necessary. Modern critics have often criticized Plato and Aristotle for this assumption, but, in his article 'Aristotle's World and Mine,' **Francis Sparshott** throws some interesting light on it. He points out that modern accounts of scientific explanation allow contingent facts to figure both in explananda and in explanantia. For example, the fact that *this* mass of 1 gram was subjected to a force of 1 dyne is thought to explain, taken together with Newton's second law of motion, why this mass was accelerated by 1 cm/sec^2, even if it was completely contingent both that the mass was subjected to this force, and that it was accelerated as a result. But Aristotle is interested in explaining only those events that conform to unchanging and repeatable patterns.

There is an interesting duality in Aristotle's conception of necessity. In the *world* necessity attaches primarily to the uncaused truths, and is inherited by everything that is necessitated by these uncaused truths. Now, as **Alan Code** says,

> Plato ha[d] no *general* account of how one gets from first principles to the theorems ... A perfectly general account of the way in which causes and principles are employed in such a way as to yield understanding must be able to answer the question: how does a systematic understander get from those truths that are known through themselves to those propositions that are known through their causes? ...
> Aristotle thinks that one knows a theorem when one sees that it *follows of necessity* from those principles that are *its* "causes." Thus ... a general characterization of the relation *follows from necessity* is mandatory.
> Furthermore, reasoning from ... principles to ... conclusions takes place in a variety of sciences ... , hence a perfectly general account of this relation must not depend on the ... particular content of the propositions of any given science, and in this sense the account will be *formal*. (128-9)

The way the problem is posed brings out clearly that there are two relationships that we have to deal with here. The first is the relationship of necessitation that obtains between cause and effect; that is a *real* relationship, in the sense that it links things *in the world*. The second relationship that we are concerned with stands proxy for the causal relationship in the mind of the understander. This is the relationship that connects the premises and conclusion of a demonstration, and as we have seen it is a species of deductive relationship. This second relationship should not be confused with the first; it does not link things in the world, but rather propositions that describe things in the world. It is through this second relationship that, in **Code's** words, 'a systematic understander get[s] from those truths that are *known* through themselves to those propositions that are *known* through their causes' (128). It is the second relationship that is formal, and it is this relationship that we are causally necessitated to grasp by the very nature of our minds.

But what is Aristotle's account of the *first* relationship, the relationship between cause and effect 'out there'? What is the connection between the formal relationships that fall under the purview of first philosophy and real causal relations? The first thing to say here is that after something like the formal opposition of contraries is studied by *first philosophy*, the principles arrived at have to be applied to particular pairs of opposed properties. A typical example of such an application occurs at *de Somniis* 453b25-454a1:

> This much is clear, that waking and sleep appertain to the same part of an animal, inasmuch as they are opposites, and sleep is evidently a privation of waking. For contraries, in natural as well as in all other matters, are seen always to present themselves in the same subject, and to be affections of the same: examples are − health and sickness, beauty and ugliness, strength and weakness, sight and blindness, hearing and deafness.

Given that sleep is identified as a privation of waking, it is concluded that what applies to all contraries applies to this pair. What applies to all contraries is discovered by a recitation of examples, in other words by induction. The same sort of pattern is found in the proof that we cannot believe contradictory propositions simultaneously (1005b22-34): first believing that p is identified as a contrary of believing that not-p, and then the principle of non-contradiction is applied to this pair (see **Code** 140-1).

This pattern of identifying something as an instance of F and then applying universal generalizations concerning Fs to the thing is a familiar part of Aristotelian scientific procedure (*Post Anal* 71a17-24). But there is something odd about these occurrences of it. Consider the syllogism:

Contraries belong to the same subject.
Sleep and waking are contraries.
Therefore, sleep and waking belong to the same subject (i.e. to the same part of an animal).

Evidently the second premiss and the conclusion of this syllogism belong to biology, since 'sleep' and 'waking' are terms that belong to that science. The first premiss, however, is an 'axiom,' and 'contrary' is a 'common notion' (the latter is Euclid's term not Aristotle's, but if axioms are to be permitted to enter a science, we need common notions). So far, so good; but what about the *causal* relationship between sleep and waking — is this a subject for study by first philosophy?

If the application of the principle of contraries to biology were straightforward, first philosophy would study this causal relationship, not in its particularity, but qua relationship between contraries. And this would mean that causal relations on the domain of a 'special science' would be studied by another science — a suspicious consequence since, as we have seen, the structure of a science is supposed to reflect the network of causal relations on its domain.

But the application is *not* straightforward: Aristotle says that the common axioms apply to the particular sciences 'by analogy,' and are to be employed only insofar as they apply to the subject-matter being discussed (*Post Anal* 76a39-40, and *Metaph* 1070a31-2). This seems to imply that the premisses in demonstrative syllogisms in the special sciences are never axioms, they are axioms restricted to the domains studied by these sciences. Now this is a directive completely at odds with the spirit of *Post Anal* I.5, which demands that universal generalizations can only be known when their subject terms are specified with full generality: a man does not know that a triangle has angles equal to the sum of two right angles, Aristotle says, if he proves this fact separately for each kind of triangle, for then he would not know that it is triangularity as such that accounts for the property, and not equilateral, isosceles, and scalene triangularity separately (74a25-32). Now, to restrict axioms to some sub-domain defined on Being is to commit just such an error: for it is only as applied to Being as a whole, that is, through first philosophy, that the axioms are known. It follows that if the first premiss of the syllogism that we are discussing, that contraries cannot belong to the same thing, is restricted to the domain of biology, it cannot be known, strictly speaking. The syllogism would not, then, be a demonstration, and its conclusion would not be known. So if the proper domain of the ontological relationships corresponding to the axioms is Being, restriction to a sub-domain cannot be how the axioms are made to operate in the special sciences.[13]

What is the significance of Aristotle's claim that the axioms apply to the special sciences only 'by analogy'? I shall start by taking Aristotle literally when he says that the axioms have to be restricted to subdomains of Being before being used by the special sciences. Let us suppose therefore that the causal relationship that holds between particular pairs of contraries — sleep and waking, heat and cold, and so on — is proprietrary to these pairs (or perhaps to the pairs of contraries found in a science, taken collectively). In other words the causal relationship between heat and cold stems from what *they* are, and is not transmitted to them from the nature of contraries. This would imply that the question with which I began this section: 'What is Aristotle's account of the relationship between cause and effect "out there"?' cannot be answered. There are too many such relationships, and each one holds properly on some sub-domain of Being. But this cannot be all, for what is it that constitutes the *analogical* unity of the principles of first philosophy?

The answer can only be that despite the causal fragmentation of the domain of Being noted in the last paragraph, certain truths about beings can be investigated over the entire domain. This is a fact about the way our minds are constituted; certain causally diverse truths present themselves as cognitively unitary. The theory of demonstration and proof serves as an example. The syllogism is a universal instrument of demonstration; that is, the relationship between premisses and conclusion of a demonstrative syllogism stands proxy in our minds for causal links that are quite diverse in their intrinsic nature since they are links in distinct domains; and this is what enables us to track quite different sorts of causal links. But in spite of the diversity of the causal relationships tracked by syllogisms, syllogisms can be studied in a unitary way, because our minds comprehend syllogistic principles in a unitary way. First philosophy is a separate science, and its principles unified, because the method by which we investigate it involves taking a detour through the way we comprehend things.

The possibility of a 'science' of metaphysics is rooted, then, in what amounts to a thesis about our cognitive capacities, and thus it is an instance of a phenomenon that **Sparshott** remarks on: Aristotle habitu-

---

13 For further arguments that support this conclusion see T.H. Irwin, 'Aristotle's Discovery of Metaphysics,' *Review of Metaphysics* **31** (1977) 210-27.

ally studies our epistemic and cognitive faculties as though they, and the match between them and real patterns of causal connection 'out there,' are an integral part of the study of nature. Even the causal priority of substance is 'reflected in the fact that elementary propositions assign predicates to subjects' (34-7). In the end, Aristotelian methodology of science comes down to this. Our proofs, and our conceptions of substances and other things as one and as existent, depend on certain epistemologically prior principles, which are studied by first philosophy, using a variety of methods, all of which amount, in one way or another, to examining the cognitive status of candidate principles. The *starting-points* of first philosophy are 'best-known,' which means that no justification of them is possible or necessary, except to 'show,' elenctically as **Code** says (144), that they cannot be disbelieved or even meaningfully disputed. In the end, that is the only warrant we need in order to adopt them.

But what sort of certificate of reliability is this? Might it not be the case that our inability to disbelieve something is merely a mark of our own limitations? Hume saw this point: he supposed that our propensity to think causally was nothing more than an innate tendency of the mind, and he concluded that our causal beliefs cannot be justified by 'reason.' Kant disagreed, but this was because he was an idealist about the propensities of the mind: space, time, and causality, as well as the 'principles of substance' (as Aristotle would call them) are merely reflections of the preconditions of perception and thought, and these concepts and categories are justifiable only as pertaining to these preconditions, but only because there is nothing that corresponds to them 'out there.' But Aristotle is a realist. Why should *he* have thought that our mind's inability to disbelieve something is a warrant for believing it?

The answer is suggested by **Sparshott**: Aristotelian science 'forms a single consistent analysis of knowledge as the replication of the system the world actually has, as if the cognitive apparatus and devices of human beings had precisely this function, to bring the world to consciousness' (36). **Jonathan Barnes'** article, 'An Aristotelian Way With Scepticism,' discusses various epistemological moves to justify what we naturally believe that can be based on such a vision. He suggests that Aristotelian epistemology is probably teleological, holding that the good of rational creatures consists in the contemplation of the true, and that the mind is constituted for this end. If it is impossible to conceive of a certain proposition as being true, then, that must be because to conceive of it as true would be detrimental to the end of contemplating the true. This is a good Aristotelian argument all right, but is it credible?

Barnes thinks, and I agree, that the only credible teleological arguments are those that can be linked to 'soteriology' (56-64). That is, we should only accept that *F* is for an end *G* only if *F*'s contribution to *G* is what accounts for the survival of an organism that does *F*. Now think about the patterns of thought that we find it impossible to violate, and on which we 'base our proofs.' Is it plausible to suppose that *these* could be false given that we have survived? I am not suggesting that everything that we do under the name of reasoning is defensible: as we all know, we are prone to certain logical errors. Affirming the consequent, for example, seems to be an error that is inextricably intertwined with the human condition. But the point is that we are far from unable to detect the error in such common fallacies: though we have a regrettable tendency, qua humans, to affirm the consequent, this is not a tendency that is irreversible, and we can indeed train ourselves out of it. Think not of these principles, however, but rather of those on which we *ultimately* base our reasonings: the principle of noncontradiction, the transitivity of entailment, the 'law of identity' (*p* implies *p*) — it is difficult to see how a creature that was able to 'think' in a way that violates *these* principles could be fit for survival.

Of course, this argument is a good bit weaker than the arguments that **Barnes** criticizes. For he is concerned mainly with teleology being used to support the view that there is some internally available, universally applicable, *criterion* of truth; and he shows that credible teleological arguments will not support this view, while the one that does support it is completely lacking in plausibility. I agree that soteriological epistemology can do nothing for those who aim to rebut scepticism; nevertheless, it can perhaps do something for first philosophy.

IV Matter

We have seen that each Aristotelian 'special science' starts from propositions that apply exclusively to the domain that is under investigation, and from the axioms, which apply analogically to that domain. This leaves no room for any inter-science connections in Aristotle's scheme of things, except for the analogical application of first philosophy to the special sciences, and the use of mathematics in the special sciences.[14] It is this autonomy of the special sciences that ensures that every provable proposition is traced back to appropriate first principles.

The autonomy requirement ensures that each science will study a large number of connected kinds of things. For example, we cannot have a science that is exclusively concerned with truths about triangles, or only with circles, for then what would we do with truths about the interactions of the two — what happens when you inscribe a triangle inside a circle? So the domain of geometry must comprehend both circles and triangles, and must have some principles that are applicable to both together. The definitions of the non-basic terms will implement the common treatment of different kinds of objects: circles and triangles are both defined using the basic notions of point, line, etc., and their interactions are explainable partly in terms of the properties of the points and lines out of which they are both constructed.

Thus there will be a degree of reductionism at work *within* each science: to some extent, truths about circles and triangles are reductively explainable in terms of parts that exist and are defined independently of wholes, and the relationships that these bear to each other. But there will be no inter-science reductive relationships at all. If matter is studied by physics and nutrition by biology, it will not be possible to establish reductive connections between matter and nutrition; that is, there will be no definitions, in the technical sense, of nutritional properties, or vice versa. Evidently, this contributes to a feature of Aristotelian science commented on by **Sparshott** (32-3): each sort of thing is described in terms appropriate to it. The redescription of mental or vital characteristics in purely physical and chemical terms which is so characteristic of modern science is completely absent from

---

14 The use of mathematics has many interesting parallels with the use of first philosophy, inasmuch as it would seem that mathematics abstracts from real causal relations. According to Aristotle, mathematics studies certain attributes of physical things that are 'separable in thought' from the changeability and motion of the physical objects of which they are attributes (*Phys* 193b23-194a12). Aristotle says that you can never prove a proposition by a science to which it belongs, except when one science is subordinate to another as optics is to geometry, or harmonics to arithmetic: these are examples of physical sciences that are subordinated to mathematical sciences by the intellectual separation just referred to (*Post Anal* 75b14-17). See also *Post Anal* 76a9-15: in the case of proving propositions of harmonics by arithmetic, there is a common element; presumably this is the attribute that is separated by thought.

Aristotle's science. So are the extraordinary insights that such redescription has brought.

This is a fact that calls for some explanation. As both **Montgomery Furth** ('Aristotle on the Unity of Form') and **Marc Cohen** ('The Credibility of Aristotle's Philosophy of Mind') show in their papers, there is a pretty close connection between what happens at the material level and what happens at the level of an organism. For example, the reproduction of animal form, as **Furth** describes it (84-5), consists of a series of motions in catamenial matter initiated by semen, which carries the form; and sight corresponds, according to **Cohen** (107), to some physiological change in the 'eye-jelly.' Isn't *some* reduction implied by these connections between levels? What then are the practical limitations imposed on Aristotelian science by the autonomy of the special sciences? And is it just contingent that such explanations as the above did not occupy a larger part of Aristotle's thinking about biology and psychology?

In order to assess the situation let us look briefly at a couple of physical phenomena and the explanations that they would receive in an Aristotelian framework. Consider first a phenomenon that received attention from Galileo and other reformers of Aristotelian mechanics back to Philoponus. Consider a sphere in free fall towards the centre of the universe, which is its natural place. The sphere is composed of two hemispheres. Are these also falling freely towards their natural place? According to Aristotelian theory, the time that a body in free fall takes to traverse a distance will be proportionate to its heaviness (*de Caelo* 273b30-274a2). It would follow that if the hemispheres could be regarded as being in free fall, they would traverse the same distance as the sphere of which they are parts in half the time as the sphere: obviously impossible. Now it is probable that Aristotle did not think of this phenomenon: his successors were quite clear that it destroyed the intuitive appeal of his theory, and if it had occurred to him, he might have thought so too. Still it can be handled quite easily within what I take to be his framework: since the hemispheres cannot fall at half the rate as the sphere of which they are part, they cannot be regarded as being in independent free fall. Either their fall must be regarded as forced, presumably by the whole sphere, or there must be some sense in whch they are not seeking the centre of the universe independently of the sphere of which they are parts.

Consider now a different phenomenon. Aristotle defines the *nature* of a thing as a source of movement within itself. The elements have natures, or are natural bodies: they move themselves towards their natural places. Living organisms too are natural: they change and move themselves volitionally, and grow and decrease in obedience to internal principles of change. A *motion* is said to be natural if it is a motion of x and its cause is the nature of x. If a motion is not natural, then it is contrary to nature (*De Caelo*, 269a10-11) — that is, it is forced. Now, the motions I execute in typing this script are natural, because they are my motions and are caused by something within me. But I am composed out of the sub-lunary elements. What happens to the elements out of which I am composed when I type? Presumably they move when I move. Is their motion natural or unnatural? It is not natural: my motions in pecking away at this keyboard are not rectilinear towards or away from the centre of the universe, and that is the only sort of motion that is natural to the elements. We come then to the same dilemma as we reached concerning the hemispheres in free fall. Either the motion of the elements out of which I am composed must be regarded as forced by me, or there must be some sense in which they are not independent natures seeking the centre or the periphery of the universe.

Now I do not know whether Aristotle ever says that a natural motion of a thing should be regarded simultaneously as an unnatural motion of its material causes: its parts and its homoiomerous matter. I do not believe that he does. But think what it would take for him *systematically* to adopt this type of solution to problems like the above — he would have to have a very comprehensive ontology of parts, and a comprehensive dynamics of force and counter-force. For *every* natural motion or change in me, there is a part that is also the (*proper*) subject of change; consequently, there has to be a dynamical interaction of the nature of the part of my nature for every such change. *We* have such such an ontology of parts, atomism. And our post-Newtonian mechanics provides us with just such a dynamics, for it is a principle of our mechanics that forces acting on a whole are just resultants of forces acting on and exerted by the parts. But Aristotle has a much more restricted ontology: one form, one thing (see **Furth** [95-6]). He just does not have enough things to have a physics of force and counter-force. And he has neither a comprehensive dynamics of force, nor the kind of force (like inertia, gravity, electromagnetic force) that arises out of matter as such and thus acts as the common currency by which all interactions are measured.

So the only way out for Aristotle is to suppose that the matter and the parts of a thing do not have independent natures. In some sense they are not acting as they would if they were independent: when the parts exist as independent entities their natures are active, but when they are only parts, their natures are dormant, or merely potentials.

Though Aristotle does insist that things are composed out of the elements, he does not compose the natures of things out of the natures of parts in the way that we as post-Newtonian atomists are accustomed to do.[15]

The fact that Aristotle analyses things into elements may beguile us into thinking that he has a chemistry, but in fact he has no such thing. His science has neither the aspiration nor the means to explain the properties of organisms in terms of the properties of elements. His forms play a role in causing a thing's natural motions that do not consist in merely knitting a thing's parts together; for the form is specified neither in terms of a configuration of material causes, nor in a vocabulary of properties appropriate to the realm of the material causes.

The necessity of giving an independent explanatory role to form is further illustrated by the interesting suggestion in **Furth's** article that Aristotle's conception of form as the cause of the unity of a substance is so strong that his otherwise sane theory of the genetic transmission of traits went badly astray under its influence (91-8). The problem for Aristotle's theory of inherited traits is to account for an off-spring's resemblance to its mother and mother's mother, etc. The problem arises because what the mother contributes to the off-spring is just matter; the form is carried by sperm. Empedocles had a suggestion that might have helped: he proposed that 'in the male and the female there is a sort of tally but the whole does not come from either one' (94) — a sort of proto-meiosis. **Furth** says that Aristotle rejected this suggestion because he could not see how form could be split and still remain the unity that forms must be (95-6).

This stricture becomes quite understandable when we recall that Aristotelian forms are *properties*: 'sundering' a form is therefore not like breaking a coin in half. The unity of a form consists, according to **Furth** (a) in some sort of coherence internal to this property, and (b) in synchronic and diachronic identity conditions for substances that possess the form (83-6). Breaking up the form, whatever that might amount to, will not preserve these features. This point dovetails with a suggestion in my article about how form breaks up the unity of matter when it informs matter, thus destroying its independent existence (**Matthen**, 175). A material substance like brick or stone is itself informed matter. The form that it has has a unity of its own. When this

---

15 It is for reasons such as these that Aristotle makes the matter of a thing paronymous in it — a bronze sphere is not bronze, he says, but brazen (*Metaph* 1033a17-20) — suggesting that the matter looses its status as substance when informed.

substance is again informed by a further form, say the form of a house, the pre-existing unity is broken up because some of the things essential to the pre-existing unity might be accidental to the unity that is a house, and some of the essential properties of the house might be inherited from accidental properties of the stone out of which it is built. But this means that the form of the stone ceases to exist, since its existence is essentially tied to its unity. It is a consequence of my suggestion that what pre-exists and survives a generated substance, a living body or a house, is not identical with anything that coexists with it. This brings me into more or less direct conflict with the claim, initially made by Bernard Williams[16] and endorsed by **Cohen** (119), that something that looks, acts, and functions very much like the body *continues to exist* even after death.

It is now clear why Aristotle's science is not more reductive in nature. A thing's nature is not a result merely of the nature of its parts, indeed the nature of the pre-existent parts does not even survive their being informed. The qualification 'pre-existent' is important, because there are some such parts, the parts which are defined by their role in a whole. Such functional parts are defined in the same science as the whole to which they belong, and they do not survive nor do they pre-exist an organism. The problem of the independent natures of such parts does not arise.) The situation with form-matter reduction is therefore quite different from that of the intra-science reductive definitions that we have encountered, where triangles and other geometrical figures are defined in terms of and inherit their properties from points and lines and the ways in which these are configured in the figures. It is not, in summation, Aristotle's unawareness of the reductive strategy that makes it occur with relative infrequency in his science: rather it is his form-based ontology, and his lack of a physics of force and counter-force.[17]

These features of Aristotle's physics make it extremely difficult to assess the contemporary credibility of Aristotle's philosophy of mind, which is what **Cohen** sets out to do. **Cohen** seeks to defend a *functionalist* interpretation of Aristotle's philosophy of mind against certain objections advanced by Myles Burnyeat. Before I say something about this issue, let me define a few terms.

---

16 Bernard Williams, 'Hylomorphism,' *Oxford Studies in Ancient Philosophy* 4 (1986)

17 A number of issues concerning a form-based ontology and science are sensitively and acutely discussed by Sarah Waterlow in *Nature, Change, and Agency in Aristotle's Physics* (Oxford: Clarendon Press 1982).

A property F is *supervenient* on a family of properties G if whenever any two things are indistinguishable with respect to G they are also indistinguishable with respect to F *as a result*. (For example, materialists hold that mental properties are supervenient on material properties.)

A property F is *multirealizable* if they are two or more distinct families of properties on which it might supervene. (For example, it may be that mental properties are supervenient not only upon the properties studied by molecular biology, but also on the properties of silicon chips that are of interest to electronic engineers.)

A property F is *functional* if it is assigned to an individual on the basis of its possessing a mechanism that gives it a particular disposition to respond to various situations, where these situations and responses might be internally characterized as states of the individual, or externally as 'stimuli' to which it is subject, or behaviour that it performs. (For example, pain might be regarded as a functional property because it is defined as what brings about aversion [response] to situations that cause trauma.)

Now to **Cohen**. First, a small quibble. The issue between Burnyeat and **Cohen** is *not* about functionalism. Burnyeat wants to say that seeing red is for Aristotle nothing more or less than being aware of red. **Cohen**, defending Sorabji,[18] thinks that seeing red is intimately tied to a physiological process, the reddening of eye-jelly. But suppose seeing red is *functionally* defined in some such way as this: a process by which a sensible subject comes to know that a red object is present by the action of the red object on his senses. This definition would be satisfied both by a non-physiological process that results in awareness of the red thing, and by one that is mediated by a physiological change brought about the action of the red object. Both interpretations are perfectly compatible with Aristotle having been a functionalist. The point of contention is different. Sorabji, defended by **Cohen**, must hold (1) that perception is multirealizable, since the formal definition of perception permits it to be realizable on quite different substrata, and (2) that perception is, actually or necessarily, physiologically realized.

---

18 Richard Sorabji, 'Body and Soul in Aristotle,' *Philosophy* **49** (1974) 63-89; reprinted in J. Barnes, M. Schofield, and R. Sorabji, eds., *Articles on Aristotle, Vol. 4: Psychology and Aesthetics* (London: Duckworth 1979) 42-64

Is this correct as an interpretation of what Aristotle was after? Everything depends on what 'realizable' means. **Cohen** himself says that 'the reddening of the eye-jelly is only the material side of perception, the matter of which the perception of red is constituted' (107). But we have seen that the matter-form doctrine is wrought with components that are at home in a physics quite different from our own. It *may* be that Aristotle would hold that two things that have different perceptions are undergoing different physiological processes, and also that two co-specific individuals experiencing the same sort of perception are undergoing the same sort of physiological process. It is certainly the case that if I am undergoing a perception of red and you of blue, then you are affected as if by a blue object, and I as if by a red object. So there is *some* internal difference, and it is not implausible to think that the difference is for Aristotle de facto physiological, if not so by definition or necessity. But this is not enough. Is it the case that two perceivers of red are so *because* of the similarity in their physiological states? As we have seen, the claim the physiology is as matter to the perception as form is not enough to support *this* conclusion, because Aristotle gives form an independent role. He is not in the materialist business of trying to account for the properties of wholes in terms of the properties of their parts and the configuration thereof; and it is precisely this programme that gives functionalism its bite. The problem for **Cohen**, therefore is to complete the argument. Either he has to base the multirealizability argument on something more than the form-matter model of realizability, or he has to modify my definitions of supervenience and multirealizability in a plausible way.

Burnyeat has problems too. He has to say why Aristotle took the physiology of perception seriously; perhaps it is true that Aristotle did not think that any physiological story was *necessary*, but he obviously thought that such a story had *some* relevance. What is the relevance of the physiological story?

There are similar problems for the sort of account of matter that I have been pushing. This account is adequate to understanding material causes where these are of the sort defined in terms of the whole. But what of material causes that are both independently existent and definable and dealt with by a different science? The elements exist independently of living organisms and are dealt with by a different science. Yet Aristotle does obviously think that truths about the elements explain some of the properties of living things. How are such explanations to be integrated into his account of the structure of his science?

Aristotle's World and Mine

Galileo's *Dialogue Concerning the Two Principal World Systems* can be seen as marking the watershed between the world-views of modern science and of medieval superstition.[1] But the systems contrasted are, in the first instance, two views of celestial mechanics: Galileo's own, and that contained in Aristotle's *de Caelo*. And if we take that expression 'the world' to stand for everything there is, it is not to the *de Caelo* that we should first look to find Aristotle's account of it. This paper presents an interpretation of Aristotle's view of the world, and then says something about the significance of Galileo's substitution.

My title is ambiguous. Do I speak of 'Aristotle's World and Mine' as I might speak to a friend about our two wives as 'your wife and mine,' or as I might speak to my wife of our one child as 'your child and mine'? For reasons that will become plain, I let that ambiguity stand. But the topic for the paper first occurred to me when I was gazing absently out of my study window. What I saw from my window was buildings, cars, roads, people, squirrels, birds, grass, trees, and clouds in the blue sky. And that was all I saw: a sample of the world we pass our lives in, the busy interface between the tedious miles of rock and iron and the empty light-years of nothing much but hydrogen. It occurred to me that it was this familiar world in its familiarity that Aristotle was concerned to explain, and that there was a way in which our science has renounced this task without quite knowing that it has done so. That is what I want to write about.

Aristotle says that all our wonderings about the world and what it contains begin with astonishment that things should be as they are; but that, when one has completed one's inquiries, one would be astonished if things were otherwise (*Metaph* A, 983a12-21). That is, to explain something is to make it seem inevitable. As Norman Campbell argued some sixty years ago, we achieve this sense of inevitability

---

1 Galileo Galilei, *Dialogue Concerning the Two Chief World Systems — Ptolemaic and Copernican*, trans. Stillman Drake, second edition (Berkeley: University of California Press 1967)

by bringing things within the compass of our own mental operations.[2] What else could one do? One has, after all, no other guide in framing hypotheses than what one can envisage as a possible law. The trouble is that one may thus reach the illusion of understanding too soon; and it was because Aristotle seemed to have led his followers into this trap that he came to be thought of as the enemy of science.

More about that later. Meanwhile, back to my window. Look at those cars. Why are they so variously and brightly coloured? Part of the explanation is the invention of the necessary pigments and resins. A more important part is that people prefer them that way. Why do they? Partly because our cars belong to our dream lives: after working hours, the wage slave, bounded in his steel walnut shell, is king of infinite space; and partly because we need to recognise our car among a thousand others at a shopping mall. But then, why are those colours applied in the form of paint? Partly because painted steel is cheaper than any rustless material of comparable strength. And so we could go on. And the whole explanation of how come cars, and how come cars are as they are, in its finest details as in its broadest outlines, is an intricate web of human purposes and technical capacities, the demands of the former interplaying with the opportunities afforded by the latter. Skills of production and skills of consumption develop together, but in the last resort it is the purposes to be served that provide the dynamism for what happens and determine its form. And that is exactly what Aristotle says (*Phys* II 2, 194b1-7).

So much for the cars; and that takes care of the buildings and other artifacts as well. But what about the trees? How come they are as they are? Well, we know for a fact how cars got to be as they are, and cars are merely supplements to the world of trees and squirrels, so Aristotle looks for the same pattern of explanation here too. But now, instead of the intricate network of human intentions and ingenuities, we explain everything about a tree by its relation to the elaborate and distinctive life-cycle of whatever kind of tree it is, as one of a self-perpetuating species of self-maintaining organisms, in an environment in and from which it sustains itself.[3] Aristotle was in no position to establish the details of any such explanation, though he made a lot

---

2 N.R. Campbell, *Physics: The Elements* (Cambridge: Cambridge University Press 1920)

3 For the self-sustaining cycle ('Man begets man'), see *Phys* II 2, 194b13, *De Anima* II 4, 416b23 and elsewhere; for nature's provision, *Pol* I 8, 1256b10-20.

of guesses; but he insisted that that was the form an explanation would have to take. The important point is that there is nothing more to know. Oaks have no reason for being oaks; they just are.[4]

Many historians, even quite eminent ones, get this wrong. They think Aristotle must be saying that trees are trying to be trees, or that at least he must be making some covert reference to plans or intentions. But such reference is limited to rhetorical flourishes.[5] The reproduction of plants and animals is presented as a self-perpetuating system in which a small component supplied by a male triggers a series of controlled reactions and developments in the material supplied by a female; and these instructions and materials, he says, have not been correctly specified unless we can see that, if they are given, the development *must* follow.[6]

At the centre of Aristotle's thought here is a study of chicken embryos at various stages of development.[7] An unfertilized egg does not develop, but stays as it is until it rots. But in an egg that a cock has fertilized—that is, supplied with a blueprint—a chicken is formed, in an unvarying sequential elaboration of interrelated parts, beginning with the heart, incorporating the nutriment that the egg also contains. The process continues outside the egg until the fowl is full grown; and then, like the unfertilized egg, it stays as it is until it disintegrates in the unvarying sequence of age, death, and decay. In the meantime, it will have laid or fertilized eggs of its own, of which just the same account is to be given. And in all of this process there is nothing to understand except the actual sequence of events by which the bird grows, and the way its parts contribute to its maintenance and its life.

It is when we see such stable sequences of events endlessly repeated within the limits of material controls and facilitations that we recognize something that could in principle be explained. And it is the

---

4 The point here is that the formal cause functions as final, not that the final cause furnishes the formal cause. Cf. *Phys* II 7, 198a25-198b9.

5 See for example *Gen An* I 23, 731a24, and II 16, 744b15, for nature's personification as rational craftsman and thrifty housekeeper. But, though these flourishes play no part in any argument, Aristotle exploits their suggestiveness in a way that is either careless or unscrupulous.

6 *Phys* II 7, 198b5—an important statement, coming as it does in the summing up of the interrelation of the 'four causes,' and one that is too seldom noticed.

7 *Hist An* VI 3. In emphasizing this passage, I am following Marjorie Grene's *Portrait of Aristotle* (Chicago: University of Chicago Press 1979).

complexities of the most developed form, the adult in its prime, that provide the measure of what it is that we have to understand.[8] The concept of deliberation or planning plays no part, either in what has to be explained or in the explanation itself. On the contrary, human purposive action, as realized in automobile manufacture and use, is a contrivance to produce *occasional* objects and events that will fit into the *regular* world of natural things and changes. Art imitates nature, not because nature is purposive in the way that we are, but because human purposefulness mimics natural regularity.[9]

Because human purpose only mimics natural regularity, Aristotle conceives of mind in a way that is the opposite of that favoured by some recent philosophers. The sign of mind, he thought, is rational and regular order, hence predictability.[10] To some philosophers nowadays, the sign of mind is intelligent spontaneity in a situation, hence unpredictability.[11] That is because Aristotle makes practical reasoning secondary to theory. The relevant contrast is between an intelligence that grasps the system of the world, and an animal that intervenes effectively within that world. To the former, the animal is a meddler; to the latter, the system is stupidly inert.

More of that later. Meanwhile, the key to Aristotle's thought is that natural processes, as described, are self-sustaining and self-repeating. They go on for ever. Most important, they have *already* gone on for ever, because there is no way the world could have got started.[12] It follows that all the processes that make up the order of nature must

---

8 This is because explanations must be complete: every aspect of the form must be accounted for. And, because the future is in one sense open, a change can only be defined by completion, recognized by the achievement of a condition already known to be stable.

9 See the end of *Phys* II, 199a20-31, where the fact of existing 'for the sake of something' is distinguished from the procedures of searching and deliberating.

10 Compare the remarks on law at *Pol* III 16, 1287a28. But the position is a staple of Platonic/Aristotelian philosophy.

11 This view of the matter goes with an emphasis on a contrast between the rational will as undetermined and the necessity of natural processes. Since Aristotle was at pains to repudiate the mechanical interpretation of necessity, his thoughts did not tend in this direction. Those of the Epicureans, by contrast, did: for the postulation of the *clinamen* as preserving free will, see John Rist, *Epicurus* (Cambridge: Cambridge University Press 1972), 92ff.

12 The theses that the world cannot have begun and cannot end are argued in *Phys* VIII 1.

be such that they can be repeated endlessly. All orbits of heavenly bodies must be uniform and circular, not because circles are pretty, but because they must end where they are to begin again.[13] All living species *must* be permanent: a shark's feeding equipment must be of limited efficiency, because otherwise its prey would have been exhausted an infinite time ago, and there would be no sharks.[14] The world can have no history (if human civilizations have histories, that shows that human history is interrupted by cataclysms after which fresh starts must be made—cf. *Meteor* I 14, 351b8ff.). And if no history is possible that means that neither the world nor anything in it can be understood in any other way than by grasping the patterns it actually has and has always manifested. The world is as the world is, and the end of our understanding would be to be unable to believe that it could be otherwise, or could be otherwise explained.[15]

Students of Aristotle may feel, at this point, that he has played a confidence trick on us. Even if Aristotle attributes no purpose to nature, and cannot do so if the world is as stable as a genuinely everlasting world has to be, we want to say that the way he understands nature

---

13 See *de Gen et Corr* II 11, 338a15, and *Phys* VIII 1. The condition (as stated in my text here) is fulfilled by elliptical orbits as well. But the point about motion in a circle is that there is never any change of direction, no point at which the curvature is different from any other, hence no point at which any change might be occasioned. An elliptical orbit is constantly changing. Hence the difficulty of accepting elliptical orbits for planets; their acceptance does in fact go with the realization that orbits change through time, and the heavens are not stable.

14 *Part An* IV 13, 696b25-32: because sharks have their mouths underneath, they must roll over to bite, which slows them down enough for their prey to have a chance of escaping—as well as preventing the shark itself from overeating. Since Aristotle defines natural things as those containing within themselves the sources of their own movements and restings, a natural thing should contain within itself its own energy source or food supply, so that the natural unit ought to be the eco-system rather than the organism. It is a matter of everyday observation, as Aristotle says, that some things are 'natural' in the way that organisms are, and that what we mean by calling them natural is this relative autonomy: what the concept of nature turns out to apply to is largely an empirical question. We do not have to postulate a kindly Mother Nature who likes to be good to sharks.

15 This explains the theory of science worked out in the *Analytics*, in which we start with an empirical generalization (the conclusion of a syllogism) and discover more and more 'middle terms' until we end with a timeless description in which every unit proposition asserts a necessary connection.

is in fact modelled on the way we *do* understand purposive action and *do not* understand anything else. That may be true, but it misses the real point of what he is saying. His real point has to do with the conditions of explanation—of scientific method, if you like. The point is that one cannot explain events until one has identified events to be explained. And an event cannot be identified until it exists—that is, until it has stopped happening.[16] Even then, one cannot be sure there is anything to explain unless the same sort of thing happens again and again; otherwise, one might be dealing with a mere chapter of accidents, like a soap opera, which one could only chronicle.

Accidents do happen. When you plant an acorn you don't know what will happen. A squirrel may dig it up, a goat may eat the sapling, a flood may uproot it, your son may back the tractor over it, anything. Only when the oak has grown up can you look back and see what it has grown into. An event has completed itself, an oak has grown. *Then* you can trace it back to its origins in form and matter. If the world can be understood at all, it is only within those limits and on those conditions. The end is primary, not because it mysteriously drags things forward from the past to the future, but precisely because only what has happened is real. What we call 'the end' is the point we recognize as the completion that defines a recognizable process, and which accordingly marks the point from which our explanations start.

That takes care of the cars and the trees, the manufactured and living things I see from my window. Now, what about the clouds?

A cloud cannot be understood in its individuality, as a car or a catalpa can. Clouds form and disperse, come and go; to understand one, we must see it as part of a larger process. The process in question is the formation and evaporation of the aqueous mass that lies between earth and air, the sea that is transformed piecemeal into water vapour, into rain and snow, into streams and lakes, and back to the sea. This moving mass of the world's water is a self-contained and relatively self-sustained substantive entity, like an animal, leaving and returning to its normal position as the sun heats it and as the sun's withdrawal lets it cool. Aristotle, like many of his predecessors, thinks of the world as made up of four such masses, earth, air, fire, and water; but the other three are less interesting. The earth just lies there hugging itself

---

16 Aristotle goes so far as to *define* motion in terms of things having stopped happening: 'The fulfillment of what exists potentially, insofar as it exists potentially, is motion' (*Phys* III 1, 201a10). He is very proud of this definition, which is indeed very subtle.

to its own centre; fire goes up towards a supposed fiery mass over-head; and air blows around in a rather disorganized way in the mid-dle. Each of these concentric masses, and their totality in their slow mutual transformations and displacements, form a self-sustaining sys-tem, and as such are 'natural,' for to be natural is simply to have with-in oneself the source of one's characteristic movements and cessations of movement (*Phys* II 2, 192b13-27). Simply to describe the way such things keep themselves going is to explain them as fully as anything can be explained.

So much for the clouds, then. What of the winds that blow them? In principle, they form part of the natural mass of air; in detail, we have to explain them by what Aristotle contrasts with nature and calls necessity. A given mass of air moves just where the adjacent bits of air happen to push it, and the clouds are carried along with it. To ex-plain such movements we would have to go back to the entire system of the turning world, in which the heat of the sun in its daily and seasonal movements round the motionless earth sets up a complex en-tirety of local movements that ultimately cause all the other movements and changes there are—an entirety which, in its detail, is obviously beyond our grasp (cf. *Phys* VIII 7). We cannot explain the specific move-ments of any given body of air, if only because such a body is arbitrar-ily demarcated. They are a mere matter of what pushes what where. There is no pattern or regularity to them. There is nothing to explain.[17] An arbitrary air mass can be explained, as the sum of the changes bring-ing about a given configuration of particles at a given time, given the specification of initial and boundary positions and velocities. But, of course, neither the initial nor the end states are ever given. What sets Aristotle apart from modern mechanics is that the latter is content with the hypothetical explanation; the applicability of this to any event in the real world is no part of science. (This theme will occupy us later.)

That takes care of the cars, the trees, the clouds, and incidentally of the ground beneath. And I have been assuming that what goes for the trees goes for the squirrels; but that isn't quite right. The squirrels, unlike the trees, which just stand there and wave, are actually doing things: chasing each other, begging for scraps, climbing the trees. Squirrels have a way of life, and it is actually the way of life that deter-

---

17 There is nothing to explain, because pushing and being pushed are two aspects of a single event (*Phys* II 3, 195b6-20). The problem of inertia, the iso-lation of which is the real origin of post-medieval mechanics, resisted solution because it was hard to see just what the problem could be.

mines what a squirrel is. They see and hear, act and react: the individual history of a tree is mostly the story of what happens to it, but the squirrel's story is what the squirrel does, and it is what it does that requires the body it has, the special shapes of teeth and tail. And, now we have disposed of the squirrels, we are left with the blue sky and the people in the streets. The sky, with its stars, is something else. Its endless orbitings, in which even the apparent waywardness of planets proves to be a regular circling of circles, follow laws of their own, exempt from the growths and decays, the accidents and mutual interferences, that render even the self-sustaining systems of the nature around us only partially and conditionally intelligible. But what about the people in the streets? They are a part of nature too. Aristotle's system includes himself. The phenomena of thought and consciousness, the looks of colours and the ringings of sounds *as we experience them*, our thoughts *in the way that we think them*, are an integral part of our world and must figure in any explanation of it. That means, in effect, that our general account of the world must be one in which the basic modes of explaining physical reality are, from the beginning, such that our own knowledge of our lives as we live them could be integrated with them.[18] It is this inclusion of the scientist, as he knows himself and as he lives his life, within the scope of his most basic explanations, that moden science renounced; and when it was reintroduced into physics at the beginning of this century it was only in the form of making the observer and his observation factors in what was observed, the 'observer' being a depersonalized locus for inspection and intervention and not a human being.[19] When Laplace told the surprised Napoleon that in his mechanical system of the world he had no need of the hypothesis of god, it was chiefly because his mechan-

---

18 This approach must be sharply contrasted with that of modern phenomenology, for which science is an abstraction within the 'life-world' and a strict philosophy begins by prescinding from judgments about reality and unreality. For Aristotle, reality is what there is, in all senses of that word. And what there is is always a substance or derived from or dependent on a substance. To be real, we must be substances too. There is no standpoint outside science from which reality could be identified.

19 This refers to the way in which popular expositions of twentieth-century physics are always written. Recent devvelopments of the methodology and philosophy of science delve more into the scope of observer bias and error, and deal with the scientist as a social being; but they do not extend to the full recognition of the scientist's inquiry as itself a natural phenomenon.

ics, unlike Newton's, really worked; but it was partly because neither his need for hypotheses nor Napoleon's surprise were any part of that system.[20] It is precisely to this renunciation of comprehensiveness that our science owes its continuously astonishing and exponentially increasing success. But it is because of that renunciation that Aristotle's way of going about things may still strike us as a more serious attempt to explain *our* world than anyone has come up with since. He got everything wrong, his system is obsolete in principle and in detail; but he saw what an explanation of the world would have to attempt.

If scientists, like other people, are part of nature, they too must be animals, like the squirrels—but, of course, not like them, because squirrels do not do science. Humans have their own way of life, as squirrels do, but their way of life is different because they use language and have a sense of times past and future: they plan their lives, have preferences and values, and organize for efficiency. And the immediate effect of this efficiency is that they have leisure—that is, their way of life allows for times and activities that are not part of the business of self-maintenance and self-reproduction to which squirrels as well as sequoias devote their time. True leisure as thus conceived can only be that very scientific activity, that understanding divorced from practical need, in which Aristotle is engaging. And so, if the scientist is part of nature, the language-use that constitutes the basis of science is also the means whereby a life of scientific activity becomes one of the possible forms of the human way of life that is what human beings really are.

Aristotle's pursuit of systematic unity can be illustrated from his theory of elements. He equates the typical forms of the four concentric masses I spoke of—earth, water, air, fire—with the basic stuffs of the world.[21] To us this seems arbitrary, and Aristotle's arguments are inadequate. He belonged to a medical clan, and he just knew, in the way

---

20 The familiar anecdote about Laplace and Napoleon is used by E. J. Dijksterhuis to seal his account of *The Mechanization of the World Picture*, trans. C. Dikshoorn (Oxford: Clarendon Press 1961), 491. In fairness to Laplace, we must remember that his topic was restricted to *celestial* mechanics—but Dijksterhuis takes no note of this.

21 *Meteor* II 1, *de Gen et Corr* II 3, *de Caelo* IV 4, etc. But it is to be noted that these play, *as elements*, a very small part in Aristotle's system: they enter into few of his explanations, directly or indirectly. Since his system works from the top down, the level of analysis they represent has little systematic significance; they are important rather as masses of stuff.

that doctors always have just *known* things, that there are these four elements.[22] But from a theoretical standpoint the important thing about these four was that they stood for the four possible ways of combining the two pairs of forces he thought he needed for what we can only call his chemistry, forces which he rather apologetically called heating and cooling, solidity and liquidity. Among these, heat is the agent of blending, and blending includes digestion, the basic process of life. As a unifying force, heat is involved in desire, the basic force in animal action. In the most advanced systems, desire takes the form of love.[23] And it turns out that it can only be a form or analogue of this same motive force of love that keeps the whole world of nature in movement, a movement which must therefore be occasioned by something that functions like a desired or loved object (*Metaph* XII 7, 1072a26, 1072b3; cf. *Phys* VIII 6 and VIII 10).

Aristotle does not state this position clearly—in fact, he derides the facile equation of heat with love as his predecessors had made it. But what does seem clear is that he was convinced that the same *sort* of force must be operative in all analogous changes in every part and aspect of reality, and that those parts and aspects include our aspirations as well as our digestions.

I spoke just now of the whole world as in movement because of its relation to a motionless object that somehow motivates it. To speak in that way is to go beyond nature and beyond science. As students of natural science, we would be wrong to make that move, and Aristotle segregates its discussion in a sort of appendix to his general theory of cosmic motion. The natural world is in fact a self-contained and self-sustaining system of systems in motion, moving in ways that will never stop because there is nothing to stop them. It is only when we reflect that we cannot understand *why* this *should* be so that we move beyond nature and, in doing so, move beyond science.[24] It is crucial to Aristotle's world view that this should be so; but the intellectual drive that

---

22 The four elements meet us for the first time in Empedocles, though it has been fashionable to pretend that they are recognizable in earlier thinkers. Aristotle tends to sneer at Empedocles, but uses a lot of his material, which may suggest that Empedocles was a spokesman for what was or became the common property of the medical science of the day.

23 *Metaph* A 4, 985a4-6; see F. Solmsen, *Aristotle's System of the Physical World* (Ithaca, NY: Cornell University Press 1960), chapter 17.

24 Cf. W. D. Ross, *Aristotle's Physics* (Oxford: Clarendon Press 1936), 99-100.

made us scientists is such that we cannot avoid that reflection.[25] And it is also crucial that this very same move beyond nature is demanded by the implications of our own activity as scientists. Since we are part of nature, the form of our knowledge is determined by the structure of the reality into which we fit as the natural knowers. We humans are that whereby the world knows itself—a standpoint which determines the form taken by Aristotle's psychology as well as his epistemology and his linguistics and logic, as well as his theory of causation and his metaphysics, which are most elaborately integrated in such a way that the structure of human knowledge can and does replicate the system of reality as it is; and this is true of the corpus of his writings as we have it, even if we prefer to think that in his personal development the components of that corpus were constructed independently at widely separated points in his career. He demonstrates (*de An* III 1) that our five senses are the only senses there could be. Each of the senses receives the world as it really is. The mind, on the other hand, can think of anything, and thus can have no determinate physical make-up of its own; it abstracts whatever is there to be abstracted (*de An* III 4). Thus as perceivers and cognizers we are cognizant of all things as they really are—in a sense, we *are* all things (*de An* III 8). All truth is rooted in present reality, by way of the causal chain that runs from object through concept to arbitrary verbal form. All reality is dependent on substances, and this dependence is reflected in the fact that elementary propositions assign predicates to subjects. Among these propositions, those in the present tense are the norm, typically recording a present fact. Contingent statements about the future cannot be either true or false, because the future is not yet real, and there is no present reality to anchor the supposed fact to— Aristotle's preferred example is that of a threatened sea-battle, of which it cannot be true either that it will or that it will not take place tomorrow, because the present reality is that the admirals are waiting to see what the weather and the omens are like before they make up their minds (*de Int* 9). Valid arguments are constructed from series of elementary propositions, each of which in the normal form is stated as the affirmation of a predicate of a subject (*Pr Anal* I 1). In the account of

---

25 The drive is that which involves us in a search for middle terms. We discover that it is not analytically true (to speak anachronistically) that the sun should be perpetually in motion. We can think of no physical reason why the sun should stop moving, or how it could; but we cannot see why it should be in motion at all, why it should not always have been a motionless ball of fire.

concept formation, we learn that concepts are formed by a sort of psychological cumulation of recognized and remembered particulars (*Post Anal* II 19); and the initial task of science is to formalize in definitions the essential constancies that make this psychological formation possible and reliable (*Post Anal* II 1-18). Finally, in the account of formal logic and its application, we discover that the task of science is the discovery of middle terms, the elimination of conjectural leaps in the system of our recognitions. And this elimination of leaps corresponds to the thesis that in the most precise causal explanation cause and effect are identical (*Phys* III 3).

When we put all this together, it forms a single consistent analysis of knowledge as the replication of the system that the world actually has, as if the cognitive apparatus and devices of human beings had precisely this function, to bring the world to consciousness. However, though the form of our knowing is determined in that way, the bare fact that we know *anything* cannot be reduced to any natural movement. The structure and validity of proofs are not affected by the sequence of steps through which we establish them. And the fact that we can have knowledge of absolutely any sort of thing shows that our cognitive powers are not coloured by any specific material or process. It is as if those powers were non-physical forces of pure knowing.[26] And, just as our plans and deliberations are laborious simulations of the inherent patternedness of natural changes, so our laborious argumentations are as if they were articulated simulacra of such knowing as a perfect mind would have, an instantaneous grasp of the

---

26 *de An* III 5. This notoriously enigmatic chapter is made to look more puzzling than it is by the conventional (medieval) chapter division; it is in fact a direct sequel to the argument that begins at 429b22. Then all one has to do is to ask oneself Aristotle's question: how, in the light of what Aristotle has already sought to establish, can thinking possibly get started? It can't come from the properties of the perceptible object, which are exhausted in sensation, in relation to the perceiving organ; it can't be the affinity of mind for intelligible forms, for the mind itself is initially formless. There must therefore be some pure form of understanding that is not relative to the physical world, a sort of pure intellectual energy. But *what* is known is determined, not by this energy, but by what there is to be known—see the preceding note. (One of the things that has led commentators to find a more determinate role for the active intellect is that they suppose that the concept-formation described in *Post Anal* II 19 is an intellectual operation—but it isn't, it is the kind of thing an animal would have to do if it were to remember and learn from experience, cf. *Metaph* I 1).

connectedness of things. But we were saying that the human capacity for knowledge cannot be less than a general ability to understand anything understandable; so the mind of which our minds are reduced versions would have to be a sort of pure understanding of intelligibility. And that, notoriously, is exactly what Aristotle says it is: a mind that is its own sole object, an insight into insight.[27]

We now reflect that the self-contained and explicable world of nature, of self-sustaining systems of which humanity is one, is a world that *lends itself to science*. It is a world fit to be understood. And how could that be, unless intelligibility itself were in some way the ground of its structure? So perhaps the metaphor whereby the quasi-chemical force of heat developed into a love for an unmoved mover of the world can be pushed a stage further. It would be untidy to postulate *two* entities beyond nature: the object of love must be the same as the ground of intelligibility.[28] We may as well call it God—everybody else does. Aristotle, unlike Laplace, does have need of this hypothesis, not within his science but to make the fact of his science intelligible. The postulation of such a being reflects the conviction that the world must be wholly explainable within a single explanatory system, of which all parts and aspects must *from the beginning* be mutually adapted to include the explainer's knowledge and feeling as well as the objects he knows and feels.

The very strength of this explanatory drive is its weakness. The demand for comprehensiveness forecloses on the admissible forms of mechanics and other specific studies, not for the sake of compatibility with other established parts of science but in the name of the possible completability of sciences yet unformed. Aristotle knew this was wrong, he insisted that every science must be built independently on its own first principles, but in practice everything he wrote had reference to a hypothetically unified world-scheme.[29] Modern science was

---

27 The argument that establishes a 'thought of thought' at *Metaph* XII 7 is essentially the same as the argument of *de An* III 5.

28 See *de An* III 10 for the argument that animal movement depends on desire for an object apprehended as desirable. Animals are not freaks, but the most fully developed and hence normal form of natural entity—if the physical world as a whole is natural, the underlying pattern of its movement should be the same as that of an animal.

29 Granted that the idea of focal meaning (πρὸς ἕν predication) enables all sciences to be unified, in a way that the *Posterior Analytics* appeared to preclude, it

to escape this trap by renouncing explanation altogether, in favour of prediction and control. I return to this theme later, but first I must say something more about Aristotle's view of humanity.

Postulating an unmoved cause of movement plays no part at all in the understanding of natural things and processes. But a non-natural aspect is built into the actuality of our own scientific thought. The epistemic structure of our knowledge is quite independent of the procedures by which we acquire it. Insofar as we are scientists we are non-natural beings. And we cannot understand human life without referring to scientific activity. (Aristotle never defined man as a 'rational animal.' He did define man as 'capable of receiving scientific knowledge,' δεκτικὸν ἐπιστῆμης [*Topics* V, 132b1, etc.], and *precisely* that is what he meant: what is distinctive in man is determined, not by his common sense or skill in inquiry or ingenuity, but by his capacity to formulate and accept scientific demonstrations.)[30] Not only is the tropism for information built into individual humans, as the success of television demonstrates, but the whole phenomenon of culture points to the institutionalization of science.[31] What is civilization but the cultivation and use of leisure? And what is leisure but emancipation from economic necessity? And what could that emancipation be for, if not to clear a space for the free use of the mind?[32] In one way or another, every advanced civilization develops in a way that testifies to this. But in Plato's and Aristotle's day, for the very first time in the world as they knew it, the systematic development of mathematics and mathematical astronomy began to reveal the real possibility of

---

does not appear that the resulting unification of science leads to any methodological reduction. It is no more than a recognition that there is only one world.

30 Note once more the transition from the last chapter of *Post Anal* II to the rest of the book—cf. note 26 above, and the associated text.

31 The universality of the human appetite for information is asserted in the opening sentence of *Metaph* I 1. Note the emphatic choice of words: *each* person is *by nature orientated to ideation*—ὀρέγονται implies an inbuilt tropism rather than a conscious choice. There is a striking difference between this and the parallel generalization that begins the *Nicomachean Ethics*, where ἐφίεσθαι δοκεῖ provides a loose generalization about conscious behavior.

32 *EN* X 7, *Metaph* I 2. See also *Pol* I 8, 1256b20-2, though I find it hard to believe that this sentence was not insinuated into our text by some over-enthusiastic Judeo-Christian commentator.

l

actually understanding the architecture of the universe.[33] The glory of the possibility intoxicated them. 'Only a god could have this privilege' the poet Simonides had written, and Aristotle quotes the line to allude to the privilege of knowing the world as it is.[34] For most of us nowadays, for whom understanding is a weary labour and for whom science is just something we are made to learn, this bewildered intoxication is incomprehensible. We suspect Plato and Aristotle of some oddball mysticism. Even the intellectuals among us cannot understand why they pretended to prefer mere contemplation to the thrills and ardours of investigation. And, of course, they didn't—not really. They knew they were not gods. We get tired, Aristotle explains.[35] But we love our inquiries for the sake of the discoveries they lead to, and for the promise of more to come. In the moment of discovery, we have an inkling of what it would be to have a perfect grasp of the world in its unity.

Any of us who have experienced in our own work, and especially if we have shared it with others, a sudden rise to a new level of understanding, can at least sympathize with the ecstasy that in this breakthrough of breakthroughs, which was nothing less than the discovery of science itself, Plato and Aristotle felt themselves entitled to feel. In any case, Aristotle certainly did feel that the impetus of cultural development and the new science combined to give intellectual life a new value, a value which by every current standard fitted Greek notions of divinity better than it fitted their notions of humanity. But he knew, of course, that most people at most times do not go in for science, and live perfectly satisfactory lives without it. So it is not surprising that among Aristotle's surviving works we find what purport to be

---

33 The significance of astronomy in the intellectual world of the fifth and fourth centuries is pointed out by Gregory Vlastos, *Plato's Universe* (Seattle: University of Washington Press 1975).

34 *Metaph*, I 2, 982b31. The word translated as 'privilege' (γέρας) is a very strong one: outside the *Laws*, Plato always uses it to refer to some special, divinely-authorized exemption from the conditions of human life, such as mortality.

35 *EN* X 6, 1176b35 etc. But note that there are no gods in the world—we are the most godlike beings there are. Hence the paradox of *EN* X 7: what is most human in us, most characteristic of our humanity, is also what is least human in us because most divine. We are essentially world-knowers and hence participate in the world's eternity in two ways, as the locus of realization of its form as well as (like other animals) by participating in the preservation of that aspect of it which is the way of life of our species (as in *De An* II 4).

two overlapping treatises on the ways of organizing human life, one of which is constructed from the first word to the last to give pride of place to the life of the mind, and the other of which, while acknowledging the theoretical primacy of intellectual activity, treats it in effect as a side issue.[36] The existence of the former version is a trial to those scholars who would like Aristotle to be a man of the world. All this stuff about contemplation, they say, must be a hangover from Aristotle's early days, when he was spellbound by that unreliable old dreamer Plato. But the fact remains that Aristotle's major treatises as we have them, on metaphysics, on natural science, on psychology, and on ethics, follow a common pattern: a development within the natural order, usually accompanied by warnings that the natural order may not be everything, ends in a statement of why it is indeed not everything. And so it must. If it is indeed true that our psychology can be explained only on the supposition that we are par excellence world-knowers, we are unique among animals in participating in two different ways in the permanent order that is the world. Like other animals, we participate by continually regenerating the species we belong to and manifesting the way of life that defines it; but we also participate by making the world order known in its stable totality. The duality is inescapable, and requires that we deal with ethics twice over, once to show what the life of happy and virtuous humans is, and once to show how intellectual life completes and transforms the picture. Modern commentators are sometimes embarrassed by the whole business, but that is because they do not really respond either to the requirements of Aristotle's world view or to his assessment of the revolution in enlightenment of which he and his older contemporaries were both witnesses and heroes.

From one point of view, then, practical life exists only to provide leisure for the intellectual life (*EE* = *EN*, 1145a8-9). But practical people do not take that point of view. They have their own standards of success and their own uses for leisure. Aristotle recognises this (*EN* X 8, 1178a9-10 et seq.). But he goes further. It is for practical people,

36 The *Eudemian Ethics* does end with a sort of apotheosis, but despite its emphasis the passage is not built into the structure of the treatise. The *EE* begins with εὐδαιμονία, the subjective condition, and stays within its limits; the *EN* begins with the objective term, the good, and inserts εὐδαιμονία into the context thus prepared. I do not suggest that Aristotle thought of the two treatises as alternatives but that, whatever their genesis, the ancient editors did well to keep both of them.

he says, to say how the intellectual life should be lived (*EN* I 2, 1093b28-94a2). What is to be studied, and to what level, is a political question. This comes as a bit of a shock, especially when we recall that in Aristotle's day no government had yet found itself bound or entitled to tell people like Aristotle what to do. Is this what the naively intellectual Aristotle used to be told over the ouzo by his old chum Philip, king of Macedon, or his father-in-law Hermeias, king of Atarneus, as they explained to him the facts of political life? I think not, for to say that something is a political matter is not to say that governments are to decide it: what governments shall decide is itself a political issue. The point being made is only that, even if theoretical issues outrank practical issues, the decision on how to deal with them is itself a practical decision: if theoreticians set their own schedules and priorities they do so, not as theorists, but as delegates of the body politic.

What shall be studied, and how, is then a matter for practical reasoning. For Aristotle, friend of kings and master of slaves, the proper business of the intellect was the appropriate fruit of the most complete leisure, a contemplative understanding of the ordered world. But for Galileo two thousands years later, citizen of a trading and manufacturing republic and himself a teacher on salary, the point was not to understand the world, but to change it. Or, more unkindly, the point was not to understand man, but to kill him. Ever since the introduction of gunpowder, the best minds had become increasingly occupied with the problem of how to aim a gun, and, hence, how to predict the flight of a cannon-ball. Even when dropping things off towers to see how fast they fell, it was often a cannon-ball they dropped. That had not been a serious problem in Aristotle's day. Swords and spears went where they were pushed, and Aristotle's dynamics is merely a theory of pushing. (There were, indeed, siege-engines worked by weights and ropes, but there was no hope of developing them into precision weapons; and there were arrows, but archers have no time to get out their slide-rules.)

Ballistics, the theory of projectiles, was one of Aristotle's weakest points. Arrows, like other things, go where they are pushed, and stop going when you stop pushing. So why does an arrow keep going after it leaves the string? Well, its motion must set up some sort of eddy in the air, mustn't it? The air displaced by the front end must somehow fill the gap left by the receding butt end. Aristotle supposes that this eddy keeps up the pressure imparted by the bowstring. As the eddy dissipates, the arrow falls.

The theory looks just plain silly to us. We feel that his mind cannot have been on what he was saying—as though he could hardly wait to get back to those chicken embryos. And I'm sure it never occurred to him that in some remote future ballistics would form the backbone of science, to the point where peace-loving philosophers in search of a paradigm for causality would have to resort to billiards as an acceptable substitute for artillery. But in fact we cannot say that Aristotle's theory of projectiles is thrown out casually—not only does he repeat it, but he uses it as a source of analogies when discussing other topics (*de Somniis* 2, 459a28-32).

The key question here is why Aristotle felt he had to say anything on the subject at all. His work is full of such theories and factual claims that he could have had no reason to think true. Why did he not confine himself to matters on which he had time to do some solid thinking and observation?

I think he had no option. It is almost impossible for us nowadays to realize how ignorant people were then on *everything* that lay outside their personal experience. Aristotle must have felt that he had to make guesses about absolutely everything, to make sure that his work on things that really interested him would fit into an overall scheme of explanation that was at least thinkable. Many of his theories look like placeholders, provisional determinations of topics for which he or someone else would eventually find time. Meanwhile, he and his associates set about almost frantically compiling an *archive*: histories, lists of dates, natural history, records of fisherman's lore and observations of all sorts.

Shortly after Galileo's day, the Royal Society of London declared its emancipation from Aristotelian authority by taking as its motto *Nullius in verba*, 'Trust no man's word.' It is easy to take this as meaning that the true scientist checks all his facts for himself. But that is silly. Scientists, like the rest of us, know very little from their own unaided investigation. We all rely on an archive, a continuously growing corpus of corroborate and organized data. Its fragility and stability alike are shown by such episodes as the fabrication of Piltdown Man, and Cyril Burt's invention of several pairs of identical twins together with the researchers who were supposed to have investigated them. Errors and frauds pass undetected for years; but sooner of later someone passes that way again, and the archive is corrected and reintegrated.

The archive, as such, seems to have been Aristotle's invention, perhaps his greatest. Plato nicknamed him 'the reader':[37] he may have

---

37 *Vita Marciana* 6. See Ingemar Düring, *Aristotle in the Ancient Biographical Tradition* (Stockholm: Almquist and Wiksel 1957), 108, for the plausibility of this assertion.

been the first man in the Greek world to realize that our knowledge of reality must rest on the compiling, storing, and recovery of data.

The great advances in science do indeed seem to coincide with improvements in compiling and distributing data: the development of research libraries just after Aristotle's time, the invention of the printing press not long before Galileo's, electronic data processors in our own. Aristotle's civilization provided him with almost nothing. It is as if he panicked. He had to do *everything*, because until one knew something of almost everything one would know almost nothing. While anything remains totally unexplained, nothing has been properly explained.

Galileo understood perfectly how Aristotle had been seduced by his dream of understanding. 'This vain presumption of understanding everything,' he makes the down-to-earth Sagredo say in his *Dialogue*, 'can have no other basis than never understanding anything' (101). But this is because of that difference I spoke of between Galileo's and Aristotle's approaches to the balance between theory and practice. For Aristotle, to understand something is to be able to explain how it fits into an intelligible world: details may be left to be filled in as one can. But Sagredo will not claim to understand anything until he can actually produce or reproduce it. However, when he can do that, he does understand it completely. 'What has knowing how to plant a grapevine in a ditch,' he asks, 'got to do with knowing how to make it take root, draw nourishment, take from this some part good for building leaves, some for forming tendrils, this for the branches, that for the grapes, the other for the skins, all this being the work of most wise nature?' (102). Since we can't synthesize vines, we can't understand biology. But we *can* shoot people with guns, we *can* build cranes and clocks; so we *can* understand mechanics. Nature, it turns out, which for Aristotle was what the scientist had to explain and the totality of what in principle he *could* explain, is just what the Galilean scientist can't explain and should not try, because the ends of nature are beyond our control. What we should concern ourselves with is production, where the end is always *our* end, and we can concentrate on what Aristotle dismissed as secondary, the means by which this limited end is brought about. The reversal is almost complete.

Let us return for a moment to my study window, to the winds that blow the clouds. For Aristotle, the movement of a given volume of air is beyond scientific explanation. It can only be referred to the overall pattern of local movements generated by the sun and transmitted throughout the world. To trace the movements in a specific volume of air we would have to know its initial condition and its boundary conditions throughout the time that concerned us; and that, even if it were possible, would be a matter of anecdote, and not of science. But it is precisely this amorphousness that makes such a volume of air a prime target for the new mechanics. For, suppose you shut the air up in a cylinder with a piston, and some tubes going in and out, then you can *control* the boundary conditions, and the very amorphousness of the air means that you can manipulate it and control the outcome. And its amorphousness makes it directly amenable to a new sort of law, apt for a science in which events are not observed but produced. The new law is hypothetical: it says, first, 'If you do *this*, *that* will happen,' and second, 'Whenever *this* happens, *that* happens.' It is not concerned with what *does* happen, which is what Aristotle wanted to know. Above all, to manipulate this obliging air you have to know not just that the air goes where it is pushed, but exactly how hard you have to push it to produce exactly how much movement. So the new science becomes predominantly a science of hypothetical relations between quantities, preoccupied with correlations and measurements.

It is because Aristotle's science is not quantitative that we find it irrelevant even when it is right. Some years ago, he was congratulated for having observed and described the 'dance of the bees' that was not rediscovered until the middle of the present century; but actually, what he says is boring, because he has nothing to say about what von Frisch discovered, the way the orientation and speed of the dance correlate with the distance, direction, and abundance of the food source that the dance discloses.[38]

Our first reaction here may be to say that Aristotle's neglect of quantitative science reflects the weakness of Greek technology. Just as he was hampered by lack of an archive, he was crippled by the absence of standard measures and measuring devices, above all of chronometers, a deficiency compounded by the lack of magnifying lenses for microscopes and telescopes. Greek jewellers seem to have used water-filled spherical lenses; but that is as far as they went. But that may be misleading. Galileo also lacked adequate measuring devices, especially timing devices. The means for conducting experiments were de-

---

38 *Hist An* IX 40, 624b7-8: 'When they reach the hive, they shake themselves off, and each is followed along by three or four others.' See Karl von Frisch, *Bees: Their Vision, Chemical Senses, and Language* (Ithaca, NY: Cornell University Press 1950). Even if Aristotle had realized the function of the movement, it would have been just one more fact about what bees do.

veloped in a hurry, because people urgently wanted to conduct them. If Galileo could derive from his experiments data adequate to decide the issue between his own and rival theories, it was not because of the previous invention of clockwork but because he figured out how to exploit the regularity of his own pulse. The difference between him and Aristotle lay not in what they could do so much as in what they aspired to do.

The difference between Aristotle's approach to science and Galileo's appears nowhere more startlingly than in their ways of establishing that space has three dimensions. Aristotle does it like this. Take a solid object. Cut it with a knife. Now you have two flat surfaces, planes. Chop again across one of the planes: now you have two flat surfaces and, between them, an edge, a straight line. Chop across the edge: now you have three flat surfaces and three edges, and where the edges meet you have a corner, a point. From now on you can chop until you are blue in the face and all you will get will be more flat surfaces, edges, and points. After the first three chops, nothing changes (*De caelo* I 1, 268a7). Galileo is inclined to deride the muddled text in which Aristotle sets this out, and himself provides a construction of great intellectual beauty, showing how, from any given point, only three lines can be drawn at right angles to each other, and argues elegantly that this is the most rational way to describe the spatial aspect of the world. Salviati, in whose mouth Galileo puts this (at the beginning of his *Dialogue*), does not notice that he and Aristotle are essaying two opposite and complementary tasks. Aristotle shows how we divide real things; Galileo shows how we construct a system of spatial coordinates.

Popular books about the modern scientific world-view often contain near the beginning a capsule description of the Aristotelian view it displaced. But they seldom say anything about the imaginative, subtle, and comprehensive scheme of the real scope of which I tried to give you some idea. E. J. Dijksterhuis starts one of the best known histories of the rise of modern science with a section on 'the scientific legacy of antiquity' which deals with mathematics, physics, chemistry, and astronomy—but omits, without even mentioning the omission, all reference to biology, let alone psychology.[39] Biology,

---

39 Dijksterhuis, *Mechanization*.... In his complaint against Aristotle's physics and chemistry (69), he takes as a paradigm of 'purely physical subjects' 'the phenomena of falling bodies and projectiles'—another of these people, apparently, who think science should be an adjunct of the armaments industry. He

apparently, is not science at all. But for Aristotle biology is not only science but holds the key to scientific method; to write as if his world view rests on his mechanics is ludicrously to misrepresent not only his own world view but the nature of the revolution that modern science represents.

In these travesty-histories, which deal only with the dismally wrong dynamics, we read all about that wretched arrow, and we are told that in Aristotle's world everything is deduced from the fact that in a circle the defining lines are radius and circumference, so that the only simple directions in a sphere are towards the middle, away from the middle, and in a circle round the middle; and that is about all we are told. In this, our modern commentators are following Galileo's own example in his *Dialogue*, which is based on a close reading of Aristotle's work on the heavenly motions (*de Caelo*) but entirely ignores the theory of the world and the theory of explanation, the context without which the theory of the heavens is arbitrary and even ridiculous. Galileo did not ridicule it, because he knew Aristotle's works and knew what he was ignoring. But his modern followers seem mostly not to know. They feel obliged to be polite about Aristotle, so as to sound broad-minded, but it is evident that they do not know what there is to be polite about, for they assume that he was trying to answer Galileo's questions and not his own. And they assume that, because they cannot imagine what other questions there could have been to ask.

Galileo knew that Aristotle was wrong about the heavens because he had seen through his telescope the spots on the sun, the mountains on the moon, and the satellites of Jupiter. But, perhaps more important, he knew Aristotle must be wrong anyway: his theory had the wrong shape. It was really two unconnected theories of motion. The account of motion below the moon, basically movement up and down, had no internal connection with the theory of motion from the moon on up, basically movement round and round. To achieve a unified astronomy, one would have to suppose the earth to spin daily on its axis and move annually round the sun. Why couldn't the ancients see that?

Actually, within a half-century of Aristotle's death if not already in his lifetime, some of them saw it perfectly well.[40] They saw it, but few

---

seems to think it is Aristotle's fault, by the way, that the work whose title we transliterate as *Physica* is not concerned with what we call physics.

40 The key figure is Aristarchus of Samos, writing about 250 BC. The key texts are assembled, with notes and translation, by Ivor Thomas, *Greek Mathematics* (London: Heinemann [Loeb Classical Library] 1941), I, 3-15.

could accept it. It was a matter of scale. They had a notion of how big the earth was, they knew that in an eclipse the earth's shadow is twice the diameter of the moon, and they thought they could see that at half-moon, when the line from moon to sun and the line from moon to astronomer made a right angle, the angle between the line from astronomer to sun and the line from astronomer to moon differed from a right angle by less than three degrees. They calculated accordingly that the sun was very big and a long way off. In fact, that was a reason for putting the sun in the middle: what business had a huge thing like that to be careering at such an excessive speed round such an inordinate daily orbit? But consider the alternative. If the earth moves round the sun in a comparably vast orbit, then from opposite ends of that orbit you ought to get a different perspective on the fixed stars. But you don't. Or, at least, *they* couldn't. And that meant that the earth's orbit, big as it was, must be of insignificant size compared with the distance of the stars. Most people just could not accept such discrepancies of size.

A striking feature of our modern world-view is the way we have come to accept, not just discrepancies of size, but discontinuities of scale. Aristotle's world was one in which speeds and sizes are continuously faster and slower, bigger and smaller, than each other. At a pinch he might have accepted that disproportionately huge, rapid, and distant sun, for so enormous an engine might have been needed to keep the whole world churning. But our own world, in which distances within solar systems, within galaxies, and between galaxies, belong to different orders of magnitude, and in which some elementary particles last for a time that bears a smaller proportion to a single breath than that breath bears to the whole history of the world till now, is a world in which Aristotle's thought-patterns have no place. The Epicureans, a little later, found no problem with it: they allowed for an infinitely large universe, and for discontinuities in scale between the world of appearance and the world revealed by science that could have been comparable to our own: but to Epicurus and his followers the world was not a single system, and its phenonmena were not in detail explicable.

Salviati, in Galileo's dialogue, is contemptuous of this mistrust of huge distances. Either you understand these big numbers or you don't, he says. If you don't, don't talk about them; but, if you do understand them, don't suppose that what is within your comprehension is beyond the capacity of an infinite Creator. It is a telling point against the Christianized Aristotelianism of Simplicio in the dialogue, but not against Aristotle himelf, whose universe contains his own understanding mind as a proper part. The paradigm of the explainable in Aristotle's world is an animal, a compact system; if the universe is to be explainable it, too, must be a compact system.

The deepest reason why Aristotle would not have liked Galileo's world is that on the latter's heliocentric hypothesis the world I see from my window, the world that Aristotle saw through the doorway of his house, has no place. In that thin rind of earth and the adjacent air and water is the animal kingdom, which is after all our kingdom; and the animal kingdom is left out of Galileo's science. Even if there are innumerable planets and countless living things on all of them, it makes no difference; those countlessly innumerable living things will be left out too.

If Aristotle erred, as Galileo rightly said he did, by splitting the world into two regions, Galileo did worse. He deliberately excluded the phenomena of life and consciousness from what had to be explained. The world system was to be a system of mechanics. It is to this singlemindedness, this exclusion of experienced quality from the domain of science, that science owes its triumphs. It has never had to look back or sideways. Nowadays we are so used to the exclusion that we take no notice of it. But in laying the foundations for the triumph of science Galileo and his contemporaries had done an extraordinary thing. They had formulated a world system in which it was impossible that they themselves could have a place. How could they bring themselves to do something so monstrously silly?

I suspect the answer is simple. 'I don't know how true it is,' says Sagredo in Galileo's *Dialogue* (272), 'that the movement of animals is natural rather than constrained. Rather it can be truly said that the soul naturally moves the members with a preternatural motion.'[41] Animals as such, in fact, are not a part of nature. This is the Judeo-Christian world of the book of Genesis, in which the heavens and earth are created first, then the plants and animals, and finally man. Man is no part of the original order; nor are animals. They are special creations. No wonder we can't understand them. Unlike Aristotle, Galileo had no need to fill the gaps of his world by desperate conjecture. He could leave everything to the unfathomed wisdom of an infinite Creator.

---

41 Actually, this could be a shrewd remark on the account of animal motion given in *de An* III 9, 433b13-30, which certainly reads much less like an account of a being naturally moving itself than of a soul mechanically operating a body.

Herbert Butterfield taught us some decades ago to think that the methods of modern science depended on Christianity and the other religions of the Book.[42] To Aristotle, the world's order is imperfect: matter resists form, and there is no reason to suppose that there are any sums for a quantitative science to get exactly right.[43] But the laws of a world created from nothing by an omnipotent being will be obeyed to the letter. The same creationist world-view has another consequence, though: that living and conscious beings may follow different laws from inanimate nature, or no laws at all. So Aristotle's task, to find an explanation of the world he lived in and of himself as living in it, is no longer a task.

The exact sciences go their triumphant way in a world that is not Aristotle's, and they are beginning to master the workings of our own bodies and brains. It is their world that we share, and it is a privilege to do so. But there is a danger that the spiritual descendants of Galileo may forget that their world-view is not a view of the whole world. It is based on deliberately ruling out of account just those phenomena that are of most concern to human beings. They may then mistake the basic principle of their science, of physics in the last resort, for the principles of an entire world-view, so that what was deliberately ruled out at the start is deemed to have been implicitly included. The task of explaining the world may then appear not as one of finding a way of unifying our qualitative knowledge of the world with the exact sciences, but of reducing our knowledge of ourselves and our world to terms of a physics the form of which was determined by the decision to exclude that knowledge from its scope. As a guide to research, such a way of looking at things could only be beneficial. What gives one pause is the assumption that such reduction must be possible, even though nothing we do has any bearing on it, and even though we have not the faintest notion of what form such a reduction might take.

A few years ago, I heard a talk given by a philosopher who has become quite famous for the brilliance and vigour with which he

---

42 Herbert Butterfield, *The Origins of Modern Science* (London: Bell 1949)

43 How can matter resist form? Are not matter and form correlative terms? Or, if we choose to allow weight to passages postulating the existence of a 'prime matter,' how can we allow it any resistant force without allowing it a form? But presumably the idea is that in any thing or event the defining form is always superimposed on a 'matter' which is already a stuff or a situation with its own form — which persists as an independent principle of change or resistance.

commits this fallacy. Pretty soon, I thought I heard him say, we will be rid of the folk-psychology that leads us to speak of intentions and meanings, and then we will all think as people should in a scientific age, speaking only of things as they are which in this case is the movements of matter and the distributions of electric charges in our brains. But wait a minute, I thought; what is this man trying to tell me? Does he really want me to think I ought not to be interested in the difference of opinion between him and his philosophical rivals, but only in the patterns of discharges in their respective cortices? But why should I be interested in those? And why would he want me to be? But then it occurred to me that I was wrong. He could not, if he was sincere, intend me to *think* in such a way; he could only be wanting to alter the flow of ions in my brain. What a funny thing to want to do. But then it occurred to me that, even if that was what he wanted, it could not be what he *wanted* to want. He must wish he didn't want anything. If only he could see things as they really are, his neurons firing away, his voicebox vibrating, his tongue flapping, like a good neoGalilean boy. But no, not even that: it is not to *see* things, even as they are, that he must be wishing. It must be ... but one cannot say what it must be. In fact, he is talking nonsense—what a good thing that the distinction between sense and nonsense holds only in the obsolete folk-psychology.

The moral is that if a world-view leaves the view-holder out, it cannot put him back in again, and a world that excludes me is obviously not the world in which I live. I shall continue to live with my feet on the floor or my seat on a chair, looking at the things and people in the world with which I interact. The conditions that a scientific account of that world must fulfil were recognized by Aristotle. My world is his, though his is not mine.

An Aristotelian Way With Scepticism

I Aristotle and Dogmatism

Whatever we know, we know either indirectly or directly. We know something indirectly if our knowledge of it is based on our knowledge of something else. We know something directly if we know it but do not know it indirectly. Epistemology therefore has two tasks: it must account for indirect knowledge and it must account for direct knowledge — it must, in an ancient jargon, supply both a theory of 'signs and proofs' and also a theory of 'the criterion of truth.'[1] Since indirect knowledge (we may plausibly assume)[2] will turn out to depend upon direct knowledge, the theory of the criterion will be prior to the theory of signs.

Dogmatists maintain that knowledge exists: they are committed to the view that satisfactory accounts of signs and proofs and of the criterion can in principle be constructed. Sceptics argue that no such accounts are — or perhaps ever can be — satisfactory: they therefore say goodbye to knowledge — and, in the best Greek cases, they find themselves in a state of ἐποχή and ἀφασία, believing nothing and stating nothing.[3]

---

1 See e.g. Sextus, *M* VII 24-5: 'Logic includes the study of criteria and of proofs. ... What is evident is thought to be known directly (αὐτόθεν), by means of a criterion; what is unclear is supposed to be tracked down on the basis of inference (μετάβασις) from what is evident, by means of signs and proofs.'

2 I say 'plausibly' rather than 'necessarily,' since there is nothing in the *definition* of indirect knowledge which entails that it rests ultimately on direct knowledge. Aristotle's arguments in *Post Anal* I 3 against 'circular' ἀπόδειξις and against infinite chains of ἀποδείξεις attempt in effect to *prove* that indirect knowledge must have direct knowledge as its ultimate ground.

3 But on the complicated and controversial question of the scope of ἐποχή in Greek scepticism see J. Barnes, 'The Beliefs of a Pyrrhonist,' *Elenchos* 4 (1983) 5-43.

Aristotle was a Dogmatist. He must therefore have subscribed, in some sense, to a theory of signs and proofs and to a theory of the criterion. For proofs and the theory of indirect knowledge we need look no further than the *Posterior Analytics*; for ἐπιστήμη ἀποδεικτική is the optimal form of indirect knowledge, and Aristotle's elaborate account of ἀπόδειξις is at any rate the one half of a theory of signs and proofs.[4] As for the criterion and the theory of direct knowledge, later philosophers had no difficulty in discovering that too in Aristotle's texts. Thus Sextus Empiricus:

> Aristotle and Theophrastus, and the Peripatetics in general, holding that the nature of things is, at the most abstract level, two-fold (since some things, as I said before, are objects of perception and others are objects of thought), themselves also maintain that the criterion is two-fold — perception for objects of perception and thought for objects of thought (and in common to both, as Theophrastus said, is the evident [τὸ ἐναργές]). Now first in order [τάξει] is the non-rational and non-demonstrable criterion, perception; but thought is first in importance [δυνάμει],[5] even if it is supposed to come second after perception in order. (*M* VII 216-18)[6]

Sextus enlarges on this unilluminating account for a few paragraphs (*M* VII 216-26); but the enlargement adds no bright detail.[7]

---

4 The other half, the theory of 'signs,' was far less well developed; but see M.F. Burnyeat, 'The Origins of Non-deductive Inference,' in J. Barnes, J. Brunschwig, M.F. Burnyeat and M. Schofield, eds., *Science and Speculation* (Cambridge 1982).

5 For an exactly analogous use of the τάξις / δύναμις contrast see Longinus, *subl* I i.

6 Cf. Stobaeus, *ecl* I lviii 1 = Arius Didymus, frag 16 D.

7 There has been much debate over the sources of Sextus' doxography in *M* VII (see most recently H. Tarrant, *Scepticism or Platonism?* [Cambridge 1985], 89-114). The favoured candidate is Antiochus of Ascalon, who is referred to at *M* VII 162 and 201-2. But there are powerful reasons against Antiochus, and no alternative candidate commends himself.

II Aristotle and Scepticism

It is often said that problems of epistemology only come centre-stage in the Hellenistic period. It is sometimes inferred that such problems were foreign to Aristotle — that it is inappropriate, or anachronistic, or perhaps merely pointless, to guess what his views on the criterion might have been: Sextus and the later doxographers, it is alleged, simply foisted their own epistemological obsessions upon Aristotle — we need not take their allegations seriously. And in any case (it may be added) Sextus himself does not seem particularly interested in Peripatetic epistemology. The Dogmatists who engross his attention are the Epicureans and — above all — the Stoics.

There is truth in all this. Certainly the epistemological manoeuvres of the Stoics are more complicated — and more interesting — than any carried out under Aristotle's generalship. But although the theory of knowledge did not reach its philosophical zenith until the Hellenistic period, it had already risen above the horizon in Aristotle's day. Sextus' remarks at *M* VII 216-26 are not pure fantasy. They are based ultimately on a celebrated Aristotelian text, namely *Posterior Analytics* B 19. Elsewhere in the *Posterior Analytics*, notably in A 3, Aristotle quite explicitly alludes to the errors of scepticism and indicates how he thinks to avoid them. *Metaphysics* Γ contains some discussion — admittedly brief and fairly unsympathetic — of notions which later squatted at the very centre of Greek scepticism. In the *de Anima* there are passages which lead, at any rate obliquely, towards the theory of the criterion.

Recent scholarship has drawn attention to these contacts between Aristotle and sceptical ideas.[8] The history of the matter remains in many respects dark; but it is reasonably clear that Aristotle was aware of the possibility of a sceptical challenge to knowledge, and that he had given some consideration to producing a Dogmatic answer to the challenge. In ascribing a 'theory of the criterion' to Aristotle Sextus is no doubt guilty of anachronism.[9] Yet it is, I think, a harmless

---

8   See e.g. A.A. Long, 'Aristotle and the History of Greek Scepticism,' in D. O'Meara, ed., *Studies in Aristotle* (Washington, DC 1981); E.Berti, 'La critica allo scetticismo nel IV libro della *Metafisica*,' in G. Giannantoni, ed., *Lo scetticismo antico* (Naples 1981); M.R. Stopper, 'Schizzi Pirroniani,' *Phronesis* **28** (1983) 266-8.

9   The word κριτήριον appears only once in the Aristotelian *corpus*, at *Metaph* 1063a3 — i.e. in the spurious Book K of the *Metaphysics* (see P. Aubenque, 'Sur

anachronism. I do not wish to maintain that Aristotle had a passionate penchant for epistemology. But I do think that in rejecting scepticism and adumbrating a Dogmatic theory of knowledge he committed himself to something which later thinkers could properly and plausibly describe as a 'theory of the criterion.'

III Perception and Truth

If perception[10] is to be a criterion, then there must be the closest connexion between perception and truth. Epicurus later held that perception could be a criterion only if all perceptions were true, and he duly laboured to defend this apparently hopeless proposition.[11] Aristotle's view was more subtle. The most explicit statement is made in the course of his account of φαντασία:

> Perception of the proper objects is true, or contains the least possible falsity. Second comes <perception> that these things hold of something, and here already it is possible to be misled — as to its being white, there is no falsity; but as to whether the white thing is this or something else, there is falsity. Third comes <perception> of the common objects — i.e. the concomitants of what holds of the things to which the proper objects belong — I mean e.g. motion and size, which hold of objects of perception — it is especially in these cases that one can be deceived in respect of perception. (de An 428b18-25)[12]

---

l'inauthenticité du livre K de la Métaphysique,' in P. Moraux and J. Wiesner, eds., *Zweifelhaftes im Corpus Aristotelicum* [Berlin 1983]). On the history of the concept of the criterion see G. Striker, Κριτήριον τῆς 'Αληθείας (Göttingen 1974).

10 In what follows I shall speak exclusively of the role of αἴσθησις as a criterion of truth: on νόησις, the other part of Sextus' 'two-fold' criterion, I have nothing to say; nor shall I try here to deal with the views of 'Theophrastus, and the Peripatetics in general.'

11 See e.g. C.C.W. Taylor, 'All Perceptions are True,' in J. Barnes, M.F. Burnyeat and M. Schofield, eds., *Doubt and Dogmatism* (Oxford 1980).

12 ἡ αἴσθησις τῶν μὲν ἰδίων ἀληθής ἐστιν ἢ ὅτι ὀλίγιστον ἔχουσα τὸ ψεῦδος. δεύτερον δὲ τοῦ συμβεβηκέναι ταῦτα. καὶ ἐνταῦθα ἤδη ἐνδέχεται διαψεύδεσθαι. ὅτι μὲν γὰρ λευκόν οὐ ψεύδεται, εἰ δὲ τοῦτο τὸ λευκὸν ἢ ἄλλο τι ψεύδεται. τρίτον δὲ τῶν κοινῶν καὶ ἑπομένων τοῖς συμβεβηκόσιν οἷς ὑπάρχει τὰ ἴδια (λέγω δ' οἷον κίνησις καὶ μέγεθος, ἃ συμβέβηκε τοῖς αἰσθητοῖς) περὶ ἃ μάλιστα ἤδη ἔστιν ἀπατηθῆναι κατὰ τὴν αἴσθησιν.

The passage is perplexing in several ways. First, the text (as it is transmitted and as I have translated it) is almost certainly corrupt, and there is no wholly satisfactory emendation.[13] Second, there is a familiar worry about the 'proper' objects of perception (i.e. objects available only to one sense modality): does Aristotle believe, as he elsewhere says, that error about them is *impossible*? or does he rather believe, as he probably means here, that it is possible but *rare*?[14] Third, the view Aristotle promotes is surprising: perhaps it is plausible that we are least prone to error with regard to proper objects; but we might surely expect that error would be less likely in the case of 'common' objects than in the case of 'incidental' objects of perception — am I not more likely to err in judging 'That's Plato' than in judging 'That's moving'? Yet Aristotle, as though stating an uncontroversial fact, makes common objects less reliable than incidental objects.[15]

For a reader whose mind is set upon the question of the criterion, these perplexities are trifling. His questions are simpler: why should

---

13 See, e.g., R.D. Hicks, *Aristotle: De Anima* (Cambridge 1907), *ad loc.*, who reports but does not follow the suggestions of Bywater (which Ross adopts in the OCT). Despite the state of the text, there is, I think, no doubt about the general drift of Aristotle's remarks. At *An* 418a8-25 he had distinguished between (1) καθ' αὐτά or direct and (2) κατὰ συμβεβηκός or indirect objects of perception, and between (1a) direct objects which are ἴδια or proper to a single sense, and (1b) direct objects which are κοινά or common to more senses than one. At 428b18-25 he arranges the objects in the order (1a), (2), (1b): we are least likely to go wrong over (1a), most likely over (1b). (Simplicius, *in An* 216.16-18, says that περὶ ἃ μάλιστα κτλ applies both to (1b) and also to (2). Ingenious but impossible.) I assume that we 'go wrong' or 'are deceived' (ἀπατηθῆναι) and that our senses 'lie' (ψεύδεσθαι) just in case things are not as they seem to be: *x* is deceived in respect of perception iff (i) it seems perceptually to *x* that P, and (ii) not-P (the tower looks round to me but it isn't, the trumpet sounds flat to you but it isn't, the water feels cold to him but it isn't).

14 For the suggestion that perception of proper objects is always true see *de An* 418a12, 427b12, 428a11; *Sens* 442b8; cf. *Metaph* 1010b2-26. The present passage is not wholly comfortable: the ἤ in ἤ ὅτι ὀλίγιστον κτλ is usually and naturally read as 'or rather,' i.e., as indicating Aristotle's preferred view; yet "ΗΔΗ ἐνδέχεται in 428b20 strongly suggests that falsity is impossible in the case of proper objects.

15 At *Insomn* 460b3-27 Aristotle states and illustrates the thesis that ῥᾳδίως ἀπατώμεθα περὶ τὰς αἰσθήσεις ἐν τοῖς πάθεσιν ὄντες — e.g. cowards, infected by fear, see the enemy everywhere. In these examples he has in mind 'incidental' objects of perception — and it is just here, no doubt, that we should expect the passions most readily to induce error.

we think that perception connects with truth in the ways Aristotle specifies? What reasons have we — what reasons does Aristotle offer us — for trusting to our senses in these particular ways — or in any ways at all? Moreover, since Aristotle explicitly allows that perception is not always veridical, how are we to distinguish between true and false perceptions? Even if we allow that most perceptions are true, how can we take perception as a criterion unless we can determine *which* perceptions are true — unless we have, in effect, a second-order criterion for discriminating among perceptions?

In our passage in *de Anima* Γ Aristotle does not, and need not, address these simpler and larger questions — he is not here concerned with the problems of scepticism. But we might hope to find an answer somewhere. Aristotle is clear and unambiguous. He makes a detailed, explicit, and relatively controversial claim. He presumably had some reason for the claim. We might expect to discover it.

IV Nature and Perception

No passage in the surviving corpus of Aristotle's writings offers an explicit defence of the claim.[16] But several passages suggest ways in which he might have begun to defend the claim, and hence ways in which we may attempt to defend the claim on his behalf. In what follows I shall potter along one of these ways: the path does not lead as far as might be wished; nor, I should stress, is it the only or even the prettiest pathway for scholars to sniff out.[17] But it is, I think, a Peripatetic pathway.

I approach crabwise. In the third of his Ten Modes of Scepticism, Sextus urges that the senses are unreliable because they disagree with one another. He briefly considers and rejects a retort to this contention. The retort is this:

---

16 It does not follow that Aristotle wrote no explicit answer — we tend to forget that most of Aristotle's works are lost. But if the lost works *did* contain an answer, it was unavailable to ancient no less than to modern scholars.

17 Another path starts from Aristotle's thesis that 'the actuality of the sensible object is one and the same with the actuality of the sense' (*de An* 425b26): the nature of perception and its objects seems somehow to guarantee the veridicality of (at least certain) perceptions.

But nature, someone will say, has made the senses commensurate with their objects. (*PH* I 98)

The content of this remark is not clear; nor does Sextus say who — if indeed anyone — had actually put it forward. There is evidence, however, that the Stoics took there to be a natural 'fit' or symmetry between the faculty of perception and its objects; and we might antecedently have surmised that the Peripatetics held a similar view.[18]

Moreover, we might antecedently have imagined that view to have been suggested by an argument of a teleological cast. The argument might have two parts; and it might look, in schematic form, somewhat like this:

(A) Men and animals are endowed by nature with perceptual faculties.
But nature does nothing in vain.
Therefore men and animals possess their perceptual faculties for some end; i.e. for some φ we perceive in order that φ.

(B) Our perceptual faculties will not ensure that φ unless they are veridical.
But nature does nothing in vain.
Therefore, our perceptual faculties are veridical.

This — which I shall refer to as 'the schematic argument' — is surely Peripatetic in its general slant and tenor. Is it genuinely Aristotelian? Did Aristotle in fact hold that 'nature has made the senses' in this way 'commensurate with their objects'?

V Teleology and the Senses

Aristotle is certainly prepared to apply teleological explanation to the phenomena of perception. Thus in answer to the question 'Why [τίνος ἕνεκα] do we have more senses than one?' he suggests that it may be

in order that [ὅπως] the accompanying and common objects (e.g. motion, size, number) may be less likely to escape our notice; for if there

---

18 For some discussion see J. Annas and J. Barnes, *The Modes of Scepticism* (Cambridge 1985), 74-5.

> were only sight — say, sight of white — they would rather have escaped
> our notice and all would seem to be the same because colour and mag-
> nitude always accompany one another. But in fact since the common
> objects also belong in other objects of perception, this makes it clear that
> each of them is something different. (*de An* 425b5-11)

The argument is curious, but its sense is clear. If two qualities, Fness
and Gness, are in fact co-extensive, then we will be in danger of failing to distinguish them unless we can perceive their presence separately. Now if Fness (colour, say) is perceptible only to vision, while Gness
(size, say) is perceptible both to vision and to touch, we may perceive
the presence of Gness (by touch) while failing to be aware of Fness.
Thus we shall not ignore the distinction between the two qualities. We
possess more senses than one, Aristotle suggests, precisely *in order that*
we may be sure of marking such distinctions.

Aristotle's text contains a second teleological argument about perception, more general in its import and more pertinent to our present
interests. The argument appears twice, first in the *de Anima* and then
again in the *de Sensu*. In *de An* Aristotle begins as follows:

> Animals must have perception if nature does nothing in vain. For all
> natural characteristics hold for the sake of something [ἕνεχά του] (or else
> are accessories of things which hold for the sake of something). Now
> if every body capable of locomotion would, if it did not possess percep-
> tion, perish and fail to reach its goal [τέλος], which is the function of na-
> ture — for how will it get its nutrition? ... (*de An* 434a30-b2)

Aristotle proceeds to argue that two of the senses, touch and taste,
are essential to an animal's survival (434b9-24), and that the other senses
are present τοῦ εὖ ἕνεχα (434b9-29).

The argument is not lucid in its details; and the *reprise* in the *de Sensu* is less cluttered:

> Each animal, *qua* animal, must possess perception; for it is by this that
> we distinguish being an animal from not being an animal. And if we
> now consider animals kind by kind, touch and taste necessarily accom-
> pany all of them — touch for the reason we gave in the *de Anima* and
> taste because of nourishment. (For it is taste which distinguishes the
> pleasant from the painful in nourishment, so that one is pursued and
> the other avoided. And in general flavour is an affection of the nutri-
> tive.) As for the senses which depend on an external medium — i.e. smell
> and hearing and sight — they belong to animals with the power of loco-

motion: all who possess these senses have them for the sake of survival [σωτηρίας ἕνεκα] — in order that [ὅπως] they may perceive in advance what is nutritious and pursue it and avoid what is bad or fatal; and those animals which also enjoy intelligence [φρόνησις] [possess these senses] for the sake of their good [τοῦ εὖ ἕνεκα] — for the senses report many differences from which arises intelligence both about objects of thought and about objects of action. (*Sens* 436b10-437a3)

All animals, by definition, possess perception. They possess perception in order — most generally — to reach their goals; or — more specifically — in order to survive; or — most specifically — in order to be able to feed. All animals, even stationary animals such as sea-anemones, possess taste and touch in order to feed. Locomotive animals possess in addition sight, hearing and smell, again in order to feed. And animals endowed with intelligence possess the distant senses for a further reason — in order to flourish or live well.[19]

For the moment let me ignore the good life and concentrate on survival. The schematic argument of the last section came first to the conclusion that for some φ animals possess perceptual faculties in order that φ. We can now see that this is a schematic generalisation of Aristotle's thesis that animals possess perceptual faculties in order that they may reach their goals, or survive, or feed.

VI Survival and Truth

So much for the first part of the schematic argument. The second part connects the value of φ with the grasp of truth. Thus we must next show that unless an animal's perceptual faculties are veracious, they will not ensure that it obtains food and so survives and reaches its goals.

Aristotle frequently observes that the senses report 'differences' in the world;[20] for in general,

> each sense is concerned with its underlying object, being present in the sense-organ *qua* sense-organ, and it discriminates the differences in its underlying object — e.g. sight discriminates between white and black,

---

19 See also *de An* 435b17-25, and frag. 48 R³ = Olympiodorus, *in Pho* IV 9 (both texts contain a few oddities); cf. *de An* 413b8-10, 414a1-3, 420b16-22.

20 See e.g. *de An* 420a10, 421a14, 422b14, 32, 435a22; *Sens* 437a3-17.

taste between sweet and bitter, and it is the same in the other cases. (*de An* 426b8-12)

At the end of the passage from the *de Sensu* which I have just quoted, Aristotle connects this 'discrimination of differences' with the τέλος or good of the animal:

> those animals which also enjoy intelligence possess these senses for the sake of their good – for [γάρ] the senses many report differences from which arises intelligence both about objects of thought and about objects of action. (*Sens* 437a1-3)

It is because (γάρ) the senses report many differences that they serve the animal's good: the veracity of the senses is a necessary condition of their attaining their goal. The connection between teleology and veridicality is thus securely made.[21]

But this connection, it might reasonably be said, is conveyed rather by a delicate hint than by a downright statement. Moreover, it is the good, and not survival, which is connected with veracity. I do not know of any passage in which Aristotle explicitly connects the veracity of perception with the survival of the perceiver; but he may very well have found the connection perfectly obvious. Surely what cows need in order to graze is a capacity to distinguish grass from stones or barbed wire or other cows. They need a discriminatory capacity, a capacity to detect real distinctions in the world. For if their perceptions did not, at least normally, discriminate correctly, they would not be able to graze: they would – unimaginably good luck apart – have long since eaten themselves to death.

The point is, I think, probably present in Cicero's *Lucullus*. The passage in question does not pretend to give Aristotle's views, but it does purport to expound doctrines which the old Dogmatists, including the Peripatetics, all embraced.[22] Cicero urges that we can never act unless our appetites are moved.

---

21 Jim Hankinson made me see the relevance of *Sens* 437a1-3 in this context.

22 The passage is part of Cicero's report of the 'Old Academic' philosophy of Antiochus of Ascalon. Antiochus claimed, notoriously, that Plato, Aristotle, and the early Stoics had all shared virtually the same views; the text I quote is therefore meant to represent the views of Aristotle *inter alios*. Antiochus' syncretism is not, I think, as silly as scholars usually suppose; but I do not, of

> But that which moves the appetite must first be perceived and be believed — and this cannot happen if what is perceived cannot be distinguished from what is false. And how can the mind be moved to desire something if it does not know whether what is perceived is appropriate or foreign to its nature?[23] (*Luc* viii 24-5)

Cicero might seem to imply that *every* belief which moves us to action must be true; but that is silly, since we are often moved to action by false beliefs, and a more charitable interpretation of the argument is possible. The core of it is this: If our perceptions did not generally distinguish truth from falsity, then they would not move us to action. For suppose that my senses have no general reliability. I am thirsty. My senses report: 'There is a tankard of beer.' Two outcomes are possible. First, I may — with trepidation — drink, or attempt to drink, the putative beer. If I follow this course, I will from time to time find myself drinking paraquat — I will very soon be dead, and my senses will not move me to action. Or second (the case which Cicero no doubt has in mind), I may stand idle. If I follow this course, my perceptions will not move me to action; nor will anything else, and I will very soon be dead from inanition.

VII Teleology, Necessity, Evolution

The argument I have just constructed (which I shall tendentiously call Aristotle's argument) runs like this:

(A*) Men and animals are endowed by nature with perceptual faculties.
But nature does nothing in vain.
Therefore, for some φ we possess perceptual faculties in order that φ — and in point of fact we possess perceptual faculties in order that we may feed and survive and attain our goals.

---

course, mean to suggest that the passage from the *Lucullus* reproduces any actual statement of Aristotle's.

23 *illud autem quod movet prius oportet videri* [= φαίνεσθαι] *eique credi; quod fieri non potest si id quod visum est discerni non poterit a falso. quo modo autem moveri animus ad adpetendum potest si id quod videtur non percipitur* [= καταλαμβάνεται] *accommodatumne naturae sit an alienum?*

(B*) Our perceptual faculties will not ensure our nourishment etc. unless they are generally veridical.
But nature does nothing in vain.
Therefore our perceptual faculties are generally veridical.

The argument is explicitly teleological: it is couched in terms of ὅπως and ἕνεκα. But here, as often, teleology can be readily 'translated' into the more acceptable language of causal necessity. And Aristotle himself suggests the translation inasmuch as his text talks indifferently of necessity and of goals and ends. Take the following passage as an example:

> The other senses — i.e. smell, sight, hearing — perceive through a medium; but if when an animal touches something it has no perception, then it will not be able [οὐ δυνήσεται] to avoid some things and grasp others — and in that case it will be impossible [ἀδύνατον] for the animal to survive. (An 434b14-18)[24]

An animal has the sense of touch *for the sake of* survival. Alternatively — and apparently equivalently — it has the sense of touch because without touch *it could not survive*.

The Aristotelian argument thus translates — or perhaps rather collapses — into what may be called the necessitarian argument:

> Animals could not survive were they not endowed with veridical perceptual faculties.
> But animals do survive.
> Therefore animals are endowed with veridical perceptual faculties.

There is a pervasive — and largely harmless — ambiguity in both the necessitarian argument and the Aristotelian argument. When these arguments refer to the survival of animals, do they mean to advert to the lifespans of *individual* animals or to the perduration of animal *species*? Is it that Hodge would not have lived so long had he not been able to discriminate mice from men? or that the kind *felis domesticus* would by now be extinct were cats' eyes congenital liars? In the former case a particular instance of the argument would look like this:

---

24 Note the several occurrences of ἀνάγχη etc. in Γ 12: 434a22, 24, 25, 27, 28, 30, b3, 9, 11, 13, 16, 17, 21, 23, 26.

Hodge lived to a decent old age.
But Hodge would not have done so had his perceptions not been generally correct.
Therefore, Hodge's perceptions were generally correct.

And so for every individual animal that survives infancy. In the latter case:

Cats have survived for thousands of years.
Hence they must often have reached maturity and reproduced.
Hence they must enjoy generally reliable perceptions.

And so for all species which have stood the test of time.

Whether individual or specific, the arguments promote a soteriological epistemology: it is our survival which explains and justifies our claims to knowledge.

The specific versions of the argument are perhaps more interesting inasmuch as they link Aristotle to the modern idea of 'evolutionary epistemology'.[25] For according to its proponents, evolutionary epistemology holds, in summary, that

among all cognitive methods possible, the one which recognizes the environment most efficiently and reliably had to be selected.[26]

Or, more fully:

Our cognitive device is a result of biological evolution. Our subjective cognitive structures match with the objective structures of the world

---

25 See e.g. D. Campbell, 'Evolutionary Epistemology,' in P. Schilpp, ed., *The Philosophy of Karl Popper* (Lasalle, IL 1973); G. Vollmer, *Evolutionäre Erkenntnistheorie* (Stuttgart 1975); A. Olding, 'Biology and Knowledge,' *Theoria* **49** (1983) 1-22; N. Tennant, 'In Defence of Evolutionary Epistemology,' *Theoria* **49** (1983) 32-48; F.M. Wuketits, ed., *Concepts and Approaches in Evolutionary Epistemology* (Dordrecht 1984). The theory likes to trace its origins to the last century; but its genuine *fons et origo* is a paper by Konrad Lorenz, 'Kants Lehre vom Apriorischen im Lichte gegenwärtiger Biologie,' *Blätter für Deutsche Philosophie* **15** (1941) 94-125 (English version in R.I. Evans, *Konrad Lorenz* [New York 1975]).

26 Rupert Riedl, quoted by F.M. Wuketits, 'Evolutionary Epistemology − a challenge to science and philosophy,' in Wuketits, *Concepts ...*, 4

because they originated by adaptation to this world. And they are partially isomorphic to this world because otherwise we could not have survived.[27]

Aristotle was, notoriously, no evolutionist: he did not believe that species *had* adapted or changed to fit their environment. But he did hold firmly to the view that species *were* adapted to their environment; for otherwise nature would have worked in vain. Thus Aristotelian species possess eternal survival value, and we should be prepared to countenance a quasi-evolutionary epistemology in Aristotle. At *de An* 434a32-b1 Aristotle claims that

> every body capable of locomotion would, if it did not possess perception, perish and fail to reach its goal [εἰς τέλος οὐκ ἂν ἔλθοι].

It is tempting[28] to connect the animal's goal or τέλος with the τέλος of its primary or 'nutritive' soul, which Aristotle has earlier described:

> Since it is right to name anything from its goal, and the goal <of the primary or nutritive soul> is to generate something like itself, the primary soul will be generative of something like itself. (*de An* 416b23-5)[29]

Unless an animal possesses veridical faculties, it will not survive to maturity — εἰς τέλος ἐλθεῖν; and unless the animal survives to maturity, it will not reproduce itself and its species will not endure. But the species does endure: the nutritive soul reaches its τέλος, since nature does nothing in vain.

But the introduction into this context of the nutritive soul with its reproductive goal may seem forced. Although the specific or 'evolutionary' version of our argument is something which Aristotle easily *could* have put forward, the texts which refer to σωτηρία plainly intend individual and not specific survival.

---

27. G.Vollmer, 'Mesocosm and Objective Knowledge,' in Wuketits, *Concepts* ..., 75

28. Themistius gave in to the temptation: *in An* 123.12-13, paraphrasing *de An* 434b1, comments that τέλος οἰκεῖον ἑκάστῳ τῶν γενητῶν ζῴων τὸ γεννῆσαι οἷον αὐτό. Contrast Philoponus, *in An* 599.15, who simply glosses εἰς τέλος by εἰς ἀκμήν; Simplicius, *in An* 319.22-6, who appears to take the τέλος to be σωτηρία.

29. Cf. *de An* 415a25-b7; *Gen An* 731b5-8, 735a17-19; *Pol* 1252a28-30.

VIII Survival Value

Soteriological epistemology is an intriguing thing. It offers a wholly empirical solution to what appears to be a thoroughly philosophical problem. The problem, in its most general form, is this: 'Do our cognitive capacities match the structure of the external world?' The answer, in its most general form, is this: 'Yes they do; for if they did not match the world we should not be here to raise the question.' In this way epistemology is 'naturalised,' or revealed as an empirical, scientific, enterprise. Of the various forms of naturalised epistemology on the market, the soteriological variety is the oldest and perhaps the least unappetising.

And these naturalised epistemologies indubitably contain a certain element of truth. Animals – individuals or species – would not be likely to survive did their discriminatory apparatus not fit them for survival; and their discriminatory apparatus would not fit them for survival unless it in some fashion matched the world in which they were obliged to survive. Nor are these truths wholly trivial; for it is at least logically possible that animals should survive – thanks to massive and continuous good fortune – even though their cognitive equipment did nothing to aid their survival.

But the truths are, for all that, epistemologically unexciting. It remains an open question whether a discriminatory apparatus which 'matches the world' in the sense of making it easier for *that* animal to live in *that* world also 'matches the world' in the sense of providing the animal with true information about the world.[30] Soteriological epistemology must suppose that any apparatus which aids survival will also provide true information. The sceptical question, then, is this: Is such a supposition justifiable?

---

30 The distinction between fitness or matching the world in the sense of having survival value and fitness or matching the world in the sense of reporting veraciously about the world is sometimes overlooked, with disastrous consequences, by evolutionary epistemologists; but it is made with all desirable clarity by e.g. Vollmer, 74-5.

IX 'Most Perceptions are True'

Before I tackle that question, I may perhaps make an obvious but not wholly unimportant remark. It is this: the argument we are considering does not, in any of its forms, lead to the extreme Epicurean conclusion that *all* perceptions are true. However plausible it may be to claim that survival depends on the reliability of the senses, it is grossly implausible to claim that animals would not survive unless their perceptions were invariably veridical. General reliability is surely enough. Indeed, general but not invariable discriminatory success seems precisely what is required to explain the generally but not invariably successful fumblings which animals make in and about the world. (In any case, no evolutionary epistemologist holds the Epicurean view;[31] and Aristotle, as we have seen, explicitly maintains that our senses are not always veridical.)

But this obvious point is something which Sceptics will seize upon. Suppose that the Aristotelian argument is sound, they will urge: the Dogmatist is still no better off. For he may know now that many — perhaps most — of his perceptions are true; but he will not know *which* of his perceptions are true. If perception is to be the criterion, it must be reliable; supposing, then, that not all perceptions are true, we must be able to determine which are true and which are not. But the Aristotelian argument will not yield this result.

A parachute manufacturer assured his customers that most of his chutes would open safely; but he could offer no way of telling — in advance of trial — which of them were safe. The pilots went to another supplier. The Aristotelian argument offers the Sceptic beliefs on similar terms: he will not buy.

It is tempting to reply on behalf of the Dogmatist that the argument does at least show that our perceptions are *probably* true. We are then part of the way to Dogmatism; for although we cannot perhaps discriminate with certainty, we can make assertions which enjoy a very high degree of probability. But this reply is sophistical. For even if it is certain that most of our perceptions are true, we cannot infer that all our perceptions are probably true[32] — and we cannot infer of any

---

31 Vollmer does indeed say that 'whenever we *make* (sense) a difference in perception there *is* a difference in reality' (74) — but he does not mean what he says.

32 Indeed, if we are sometimes victims of Pyrrhonian ἀντιθέσεις (so that *x* appears

given perception that it is probably true. We can only assert of a given perception that it is probably true if we have some way of discriminating among perceptions in terms of their probability or verisimilitude. Thus the Dogmatist cannot retreat to probabilistic assertions: such assertions presuppose the ability to discriminate, which, according to the Sceptic, the Dogmatist has yet to vindicate for himself.

Both the Sceptic, who demands a criterion of truth, and the Dogmatist, who pretends to supply one, will be dissatisfied with the Aristotelian argument. For as a supplement to the conclusion that most perceptions are true, they require a method of discriminating the many true from the few false perceptions. (Or, equivalently, they require a conclusion of the form: Every perception of type T is veridical.[33]) The Aristotelian argument, as I have presented it, does not produce such a method, and I am not able to dream up any modification of the argument which will do so. The argument may still be sound, and it may still be valuable; but it will certainly need supplementation from some other quarter if it is to satisfy the Sceptic — or the Dogmatist.

X Useful Lies

But is the argument sound and valuable? Or, to return to the question I posed earlier, does survival value imply or depend upon veridicality?

Consider the instruments on a motor-car dashboard — the speedometer, say, and the petrol gauge. In order for the motorist to survive (I mean: to survive *qua* driver, to be able to continue driving successfully), he needs these instruments: the instruments have considerable survival value, they are fitting and appropriate to his enterprise. (The instruments are not, strictly speaking, indispensable: motorists can continue driving when their instruments fail, and they could have driven had the instruments not been invented. Nonetheless, the instruments possess survival value: they make successful driving far easier to accomplish, and they make it far more probable that a motorist will be able to drive successfully.)

F in situation S, F* in S*, and there is nothing to choose between the two conflicting appearances), then we can be certain that *not* all our perceptions are probably true. (On ἀντιθέσεις see Annas and Barnes, 22-5.)

33 For example, the Aristotelian thesis that perception of proper objects is always true.

The instruments also give information — or purport to give information — about how things really are. They purport to show how fast the car is going, and how much fuel is left in the tank. Then does their survival value depend upon or imply their veracity? Is the only useful instrument a truthful instrument?

It is plain, I suppose, that useful instruments need not be particularly accurate. Old-fashioned speedometers used to tremble like interview candidates — they gave at best a flickering answer to the motorist's questions. 'How fast am I driving?' — 'Oh, about 35 or 40 m.p.h.' But inaccuracy is not the same as falsity; and although the speedometers gave approximate answers, the answers they gave were (I suppose) true: the needle flickered between 35 and 40 when and only when the car was travelling at something between 35 and 40 m.p.h.

Petrol gauges are, like speedometers, inaccurate — that is true even of the gauges on the latest formula one racing cars. Yet they (or some of them) are also, unlike speedometers, persistent liars; for — or so I am told — they usually indicate less petrol than actually remains in the tank. (Perhaps they are designed to do so.) Thus a petrol gauge will regularly misreport the state of the world: if it says that the tank contains n gallons, then the tank in fact contains, let us say, $n + 1/2$ gallons. (Except when $n = 0$.)

Yet such false gauges are wholly compatible with motoring success — they can promote and facilitate good motoring despite their lack of veracity. Nor do I mean that an intelligent motorist might learn to compensate for the fault in the gauge, tacitly adding half a gallon to the gauge's readings — for in that case he would have, in effect, an accurate and veridical gauge. I mean that the motorist may survive with a gauge which, quite unknown to him, is systematically mendacious.

And there is a further twist to the story. The motorist may actually be better off — more likely to survive — with a false gauge than he would be with a true gauge. Suppose that motorists are habitually optimistic — they tend to think they get more m.p.g. from their cars than in fact they do. Then a systematically mendacious gauge is just what they need. With a true gauge, a motorist will frequently run out of petrol in mid-journey, wrongly supposing that the half gallon which his gauge truly reports will suffice to carry him the thirty miles back home. With a false gauge, he will usually get home — though not for the reasons he supposes. For the falsity of the gauge balances the optimism of his judgments, and the balance leads to carefree motoring.

XI Evolution and Scepticism

Such cases, whether real or fantastical, prove that there is no logical connection between survival value and veracity. It is perfectly possible that a mendacious apparatus should be highly advantageous from the point of view of survival. And some evolutionists in fact suppose that our own cognitive apparatus may well be thoroughly mendacious. After all, according to the biologists,

> our brains, however pliable they may be, in the end constrain us to think in certain ways rather than in others, with no guarantee that how we think bears any relation at all to the truth.[34]

Moreover,

> if the theory of natural selection is correct, then it is very highly improbable that truth seeking, and finding, organisms − as we believe ourselves to be − would evolve.[35]

And further reflection on the demands and requirements of survival suggests a sceptical conclusion:

> Complete veridicality is probably not, in evolutionary terms, cost-effective. Organisms that must act to survive must process information. They must do so reasonably well, and reasonably fast. Quick computing cuts corners.[36]

An example is often cited: we possess colour-vision; we see things as being of different shades and hues; and we are thereby the better able to discriminate − ripe and edible tomatoes are distinguished by our colour vision from green and inedible tomatoes, and the distinction has some survival value. Colour vision is not a *sine qua non* of survival, but it helps. And yet colour vision is entirely mendacious: real things are not coloured at all; ripe tomatoes are not really red − nor are they pink or orange or blue or violet. 'Science has shown us' that

---

34 Olding, 1

35 Ibid., 11

36 Tennant, 33

colours, like other secondary qualities, are illusory, that they are no part of the real external world. Mendacity, at the greengrocer's no less than in the motor-car, has its advantages. (I cite this argument but I do not endorse it: it is in fact endorsed by many evolutionary epistemologists.)

Naturalised epistemologies can thus lead not to Dogmatism but rather to a sort of Scepticism. The particular sceptical turn I have just described is modern, but the general idea it relies upon is ancient. Sextus more than once appeals to the findings of contemporary biology in the course of his sceptical arguments. And on one occasion he explicitly suggests that since our thoughts are constructed in or by a physical body they are likely to be contaminated by its peculiar features and characteristics:

> And no doubt < the intellect > too produces some admixture of its own to add to what is announced by the senses; for we observe the existence of certain humours round each of the regions in which the Dogmatists think the "ruling part" is located — in the brain or in the heart or in whatever part of the animal one wants to locate it. (*PH* I 128)[37]

The biological circumstances and surroundings of our cognitive operations are perhaps as likely to induce error as to conserve truth.

XII Perception and Action

Such sceptical conclusions have been resisted. No doubt survival does not *entail* veracity; nevertheless (it is urged) in the case of animal perception, survival value cannot in fact be divorced from veracity. For an animal's survival is determined by its behaviour, its behaviour is determined by its choices, and its choices are determined by its perceptions. The dog survives because he gnaws at bones and not at rusty nails. He gnaws the bone after selecting it. He selects it after judging it to be nutritious. He judges it nutritious because he perceives it as nutritious.

If you are unwilling to credit dogs with so much savvy, take a human example: the point is the same. We human animals normally choose something as salutary or nutritious or otherwise advantageous

---

37 Cf. Annas and Barnes, 117-18.

just in case we perceive it as salutary or nutritious or otherwise advantageous. Now if we are to survive, most of our choices must be correct (we must normally choose x as nutritious just in case x is in fact nutritious); hence if we are to survive, most of our perceptions must be correct (we must normally perceive x as nutritious just in case x is in fact nutritious). Here, at least, systematic error is impossible: the close connection between perception and action is too tight, it allows no play.

Now a sceptic has, I suppose, at least two lines of objection to this argument. First, he will observe that the argument falsely assumes that we *perceive* things as salutary and nutritious and the like. We do not. My choice of foodstuffs may depend on my judgments as to what is nutritious, and perhaps these judgments – or most of them – must be true. But nothing yet follows about perception; for the judgments are not perceptual judgments.

'But surely,' it will be said, 'my judgments about nutritiousness will at least be based on perceptual judgments? And in that case the basic judgments must themselves be, for the most part, true, given that the judgments about nutritiousness are so.' I grant the premiss, which is plausible. I suppose, that is to say, that we are thinking of such simple cases as these: 'Those berries should be edible – they're purple'; 'Don't eat that potato – it's gone green'; 'The water should be drinkable since it isn't at all brackish to the taste.' Our judgments here are indeed based on our perceptions. But if our judgments about the edibility of berries are regularly right and also are regularly based on our judgments about the colour of berries, may we infer that our colour perceptions are also regularly veridical? Plainly not: we could be perfectly and always wrong in our ascription of colour to berries, and yet regularly correct in our judgments about their nutritive powers.

For what the nutritive facts require – as the Greek Sceptics in effect say – is a correlation between the nutritive powers of berries and the way berries *seem* to me.[38] The state of affairs is, or may be, this:

---

38 The Pyrrhonist has a 'criterion of action,' namely 'the φαινόμενα': *PH* I 21-2. There are, in other words, regular correlations between the way x seems to the Pyrrhonist and the way the Pyrrhonist treats x: y treats x in manner M iff x appears F to y. (Or better: 'Normally, y treats x ....') These correlations are all that we need to account for the Pyrrhonist's behaviour – we do not need, in addition, to ascribe beliefs to him. (Although he avows that x appears F to him, he does not believe that x is really F.) If the Pyrrhonist survives, that is simply because manner M is an appropriate way for him to treat x when x appears F to him.

When I say 'This is good for you,' I am usually right. Usually when I say 'This is good for you,' I say it on the basis of a perception — 'It looks just like a glass of Guinness.' All that follows from this is that when something looks like a glass of Guinness, then it is, generally at least, good for you. It plainly does not follow that when something looks like a glass of Guinness then it generally is a glass of Guinness. If my survival depends on my judgments and my judgments depend on my perceptions, then there must be a good correlation between how things appear to me and how they affect my health. Truth does not — or need not — enter the story.[39]

The Sceptic also has a second line of objection. How good, he may ask, must the correlation really be? And he will think here of the petrol gauge. Let us forget about the first objection and imagine that there is a direct connection between my perceptions and the actions on which my life depends. Must these perceptions be generally true? To put the point in a different way: If you were designing a survival machine, would you programme its discriminatory apparatus to report the truth? It seems clear that truth would be, in certain contexts, an important aim; but it would be only one among several aims. A good designer would also ensure that, for example, the machine produced *quick* reports (delay may spell danger), and that it produced *decisive* reports (dithering may paralyse, hedging may hinder), and that it produced *cheap* reports (energy spent on gathering information cannot be spent on gathering food); and these desiderata will compete with the demands of truth. Again, truth will be important only in certain areas: it will matter, for example, that the machine should not get false reports of the form 'That's not dangerous' — for falsity here will not long be survived. But it will not matter, or not to the same degree, if the machine gets false reports of the form 'That's dangerous' — it will then be a cowardly machine, but cowards have a notorious tendency to survive.[40] And of course, in the case of most reports, which do not bear upon survival, it will not much matter whether they are true or false.

---

39 We may, of course, have excellent reasons for rejecting scepticism and for holding that berries generally are F when they look F. Here I am not arguing for scepticism: I am simply considering one argument against scepticism, the soteriological argument. And I am suggesting that this argument, by itself, does not establish the general veracity of our perceptual apparatus.

40 The machine must not *invariably* get the report 'That's dangerous' — in that case it would never feed and so could not survive. You will cross the road safely provided that (i) whenever it is dangerous the light is at red, and (ii) at

If this is right, truth does indeed possess survival value; but it is not the only aspect of information that is valuable, and it is not always valuable. The soteriological argument, then, can show very little: it surely does not show that, since we survive, our perceptions are generally veracious. Aristotle himself seems to have a glimpse of this:

> Men have a poor sense of smell; and they smell no smellable objects unless these are accompanied by pain or pleasure — the sense-organ is not accurate. And it is reasonable to suppose that animals with hard eyes perceive colours in this way, and that the differences among colours are not detectable by them except insofar as they inspire fear or confidence. (*de An* 421a10-15)

The accuracy of our senses is, in some cases at least, measured by our needs; and our needs may not demand a particularly high degree of sophistication.[41]

XIII Perception and the Good

Up to this point I have thought only of survival: it is time to turn briefly to the good life. For Aristotle explicitly mentions the good, τὸ εὖ, as one of the ends for the sake of which animals possess perception — and if survival fails to ensure the truth of perception perhaps the good will do so.

The good of an animal is something distinct from its survival: it cannot enjoy the good life unless it survives, but it can survive without enjoying the good life. Different animals no doubt have different goods;

---

any other time the light varies randomly between red and green. (I assume that you obey the lights.)

41 See Vollmer, 80: 'A *total* isomorphism [between subjective cognition and objective structure] is neither needed nor possible. But we cannot predict from evolutionary principles alone the extent of this isomorphism. It might be very good or rather poor'; also E. Oeser, 'The Evolution of Scientific Method,' in Wuketits, *Concepts* ..., 151: '... the phylogenetically developed perceiving apparatus ... is constructed in such a way that it functions reliably only in the field of life-preservation. It does not function reliably in other fields. There it may even turn out an impediment to knowledge or a source of error' (cf. Vollmer, 85-6). With such apologists, evolutionary epistemology needs no prosecutor.

but perception, or rather the senses of sight and smell and hearing, contribute to the good life at least of those animals which have intelligence or φρόνησις. For

> those animals which also enjoy intelligence possess these senses for the sake of their good — for the senses report many differences from which arises intelligence both about objects of thought and about objects of action. (*Sens* 437a1-3)

Aristotle refers not merely to practical knowledge but also to theoretical knowledge. And it is easy to construct from this reference a teleological argument of the sort we are looking for.

The good of rational creatures consists in the exercise of their reason. The highest good consists in the exercise of theoretical reason, in θεωρία. Θεωρία is the contemplation of truths. Hence the good cannot be attained unless truths are available to the contemplator. Now any truths that mortal minds may contemplate are obtained, directly or indirectly, by way of the five senses; and it is highly plausible (to say no more) to suppose that the objects of the mind's contemplation will be true only if the perceptual reports from which they were somehow obtained are also true. Thus rational creatures like us cannot achieve τὸ εὖ unless their senses are veridical. But nature does nothing in vain. Hence the senses are veridical.

The elements of this argument are all utterly familiar and will be immediately recognised as Aristotelian. It would be a simple — and a tedious — matter to develop the argument in glowing detail and to trick it out with quotations and references.

The argument is teleological, like its earlier soteriological sibling. But unlike its sibling, it cannot be readily 'translated' into necessitarian terms. Or rather, if it were so translated it would be wholly lacking in probative force. ('Unless our senses are veridical we cannot contemplate truths. But we do contemplate truths ...' No Sceptic will regard that as a tough objection to scepticism.) The argument, then, is irreducibly teleological: it depends on the thesis that there are, given in nature, certain specifiable ends or goals for rational creatures, and that these ends or goals are regularly attained (or at least are naturally attainable). It is fashionable to speak well of Aristotle's teleology; but I do not think that anyone will now defend a teleology strong enough to support the antisceptic argument I have just sketched.

And yet the argument is, I suspect, Aristotle's ultimate answer to the Sceptic. It is not an answer which, so far as we know, Aristotle ever explicitly gave. But it is an answer constructed, in a straightforward fashion, from propositions which are expressly asserted in the Aristotelian *corpus*. Moreover, these propositions are not casual asides. They are part of the hard centre of Aristotelian philosophy.

XIV Unpopular Postscript

'Aristotle Today' was the title of the conference for which this paper was written, and contributors were invited to dwell on the relevance of Aristotle to contemporary philosophical issues. (The reader will no doubt decide for himself with what degree of good faith the invitations were severally accepted.) Where, then, is Aristotle today?

Scepticism is still with us, in one form or another. Reputable philosophers write books about it, one or two every year. Most of these reputable philosophers have, of course, read their Aristotle; but few of them have studied Aristotle — and, in particular, few have studied Aristotle's known or conjectured views with regard to scepticism. Should they have done so? Would their books have been better — richer, deeper, closer to the truth — had they done so?

It seems to me evident that the answer to this question is in almost all cases: No. Aristotle's remarks on scepticism were — as we should expect — written within a philosophical framework which we no longer regard as true or useful: the foundations of his philosophy are long exploded. We should not imagine that Aristotle's answers to scepticism will be our answers — or even that they will help us towards our answers or keep us safe from temptingly naughty answers.

The point will, perhaps, be granted — and explained by appeal to the relative insubstantiality of Aristotle's remarks on scepticism. Elsewhere and on other topics, it will be claimed, the study of Aristotle is indispensable to philosophical survival. But this claim too, I think, is certainly false — indeed, it sets things topsy-turvy. For as a matter of fact the study of Aristotle, far from aiding modern philosophers in their task of promoting modern philosophy, usually distracts them from their proper enquiries and seduces them into other operations. Half a century ago a young philosopher wanted to make a study of the theory of categories. He thought it necessary first to read Kant — and Aristotle. And he thought it necessary first to study Aristotle. Fifty years later he had achieved a scholarly eminence that few could rival. And he had written nothing about the theory of categories.

If a philosopher is interested in solving philosophical problems or in making philosophical progress, then I urge him strongly not to study

Aristotle — or any of the other dead heroes of the subject.[41] The study of Aristotle is delicious and demanding and devouring. It should not be undertaken in the vain and frivolous expectation of procuring philosophical enlightenment.[42]

---

42 Since this remark has been provokingly misconstrued, I add a tedious footnote: I neither state nor imply that philosophers should not study Aristotle; I neither state nor imply that philosophers who are primarily concerned with modern philosophical problems should not read Aristotle; I neither state nor imply that reading Aristotle will never offer such a philosopher inspiration — or solace.

43 Julia Annas saw an early draft of this paper and made a number of acute comments. Many of the participants at the Edmonton Colloquium offered useful criticisms or raised pertinent questions. An audience at McMaster University provided further stimulation to improvement. I am especially grateful to Mohan Matthen: in Edmonton we had many conversations on this topic (and others) which (to me at least) were as enjoyable as they were instructive.

Aristotle on the Unity of Form[1]

I Beings and the Being Of, Substances and the Substance Of

Aristotle opens *Metaphysics* Z with the celebrated maxim, τὸ ὂν λέγεται πολλαχῶς, 'being is said in many ways,' and his project of determining how many and in what manner interrelated these ways may be, digging beneath the surface grammar of ὄν or 'being' for the varied and complex structures in reality which (according to him) the term tends to conceal or conflate, is a central preoccupation of his thought.

One of the most evident ambiguities in the Greek expression τὸ ὄν, which is paralleled in the English word *being*, is between (A) an application to *that which is* and (B) to the being *of* that which is. In the first case, the expression can be cast in the plural ὄντα, *the beings* or *the things that are*, and is unmistakable; but it need not be so cast, and the singular is dangerously bivalent.

This distinction is of course not that between existential and predicative being, but cuts across that; it divides both (A) the existent from (B) the existence of the existent, and (A) the healthies, the whites and the braves from (B) their being white and healthy and brave.

Historically, the imperativeness of observing it was first brought out, in effect, by Parmenides, who showed, in effect, that the consequence of assuming that all there is to *the being of the things that are* is just *those*

---

1 © 1986 Montgomery Furth. This paper is a patchwork from my *Substance, Form and Psyche: An Aristotelean Metaphysics* (hereafter, *SFP*, forthcoming 1987 from Cambridge University Press). A version of it has also appeared in Volume II of the *Proceedings of the Boston Area Colloquium in Ancient Philosophy*, for 1985-86. I am indebted to the audiences at Boston College and Edmonton, and especially to Jennifer Whiting, my commentator at BC, for criticism and clarification. Earlier versions benefited from discussion at the University of Arkansas (Matchette Lecture), November 1984, and at the University of Colorado, April 1985.

things that are, i.e., that all there is to (B) is (A) (together with a few other assumptions that are largely innocent), was the monstrous and absurd doctrine of Eleatic Monism. (The most important lemma in the reasoning was to show that on this footing, *not-being is impossible*.)[2] This was enough to provoke the post-Parmenideans into analyses of 'the being *of*' that, in effect, avoided the misbegotten identification. The point received its first explicit methodological (or metatheoretical) recognition from Plato in the *Sophist* (242b-245d), although accompanied by the mischievous and quite untrue insinuation that all the systems prior to the Stranger's analysis had been oblivious to it, and had confused *being* (B) with *the beings* (A).

In Aristotle, exactly the same ambivalence holds for 'substance,' i.e., οὐσία: there is (A) the sense in which Socrates is *a* substance and Coriscus another one, the sense made obvious in talk in the plural of 'the substances,' and then there is (B) 'the substance *of*' the substances, which is elaborated in causal terms as that which *makes* a substance in the first sense a substance, and answers the question *why* it is such. In the *Categories*, substances (A) are called 'primary substances' and substances (B) are called 'secondary substances'; in *Metaphysics* Z, substances (B) are called 'primary substances' and substances (A) are called 'composites.'[3]

II Substance, Essence, and Form

As we all know, *Metaphysics* Z-H is an inquiry into the nature of substance in sense (B). What its doctrine is is still controversial after twenty-three centuries of discussion, and I do not expect to settle many of the still-outstanding questions here. But it will help in following the thread I do hope to trace, if we can be reminded of some parts of the story on which there can perhaps be some measure of agreement. We will get to plenty that is debatable soon enough.

---

2 For the story see 'Elements of Eleatic Ontology,' *Journal of the History of Philosophy* **6** (1968) 111-32, reprinted in A.P.D. Mourelatos, ed., *The Pre-Socratics: A Collection of Critical Essays* (Garden City, NY: Doubleday 1974) 241-70.

3 Cf., e.g., Z 4, 1030a10, Z 6, 1031b14, 1032a5, Z 7, 1032b1-2, Z 8, 1033b16-18, Z 11, 1037a28, b3-4, Z 17, 1041b27.

First, after Z 3 it seems to be pre-analytically assumed (it is not actually argued) that the candidate category for substance *of* a thing will be a Kind, such as Man or Dolphin or Crane, and the argument is over the correct analysis, metaphysical sorting, and internal structure of a Kind that is to serve in that office. A second assumption is that each Kind *is* in fact open to analysis, *has* an internal structure, which is intelligible and subject to rational scientific investigation. Aristotle's standard name for such an analysis of a Kind is 'definition'; and as a matter of technical terminology, standardly for Aristotle the 'essence' of anything (the τί ἦν εἶναι) is *whatever is articulated by the 'definition' of that thing*; this understanding evidently goes back very early, since it is already a basic ground-rule in the *Topics*.[4] Third, it is settled Aristotelian doctrine that 'there is no definition of individuals.'

From these it naturally follows that there are no essences *of* individuals, i.e., individuals do not *have* essences. Can this be right? Yes, on the right understanding of 'of' and 'have' — there is no definition of Socrates, and hence no essence of Socrates that he 'has' all to himself; on the other hand he certainly *is* thought to 'have' those essential properties that he cannot cease to 'have' without perishing, so on another construal he may be said to 'have' an essence after all. The problem is not merely terminological, as becomes all too apparent in *Metaphysics* Z 6.

There are three titles to deal with here: (1) Man, which is what gets the definition, (2) the essence of Man, which is what the definition articulates, and (3) Socrates the individual man, who is not definable but who must *satisfy* the definition that (1) gets, whatever that is (on pain of perishing, indeed).

Aristotle uses the expression 'essence of X' (i.e., of course, the Greek for this) *both for a relation between (2) and (1), and for a relation between (2) and (3).* It is clear that these are not the same relation. For (1) and (2) stand in the relation of identity, in particular in the relation of *a specific kind* to *itself as definitionally analyzed*. Whereas, the relationship of (2) to (3) certainly is not identity; its *Categories* antecedent is the 'predicability of the (specific) definition' of the substantial individual, and its *Metaphysics* working-out is highly complex. I will have more to say of it later. For now, it is enough to see the difference between the sense of 'having' and 'of' in which the kind Man 'has'$_{1-2}$ an essence,

---

4 E.g. *Topics* I 4, 101b21, I 5, 101b38, I 8, 103b10, VI 1, 139a33 (and VI passim), VII 3, 153a15, VII 5, 154a31. *Metaphysics*: Z 4, 1030a7, Z 5, 1031a12, H 1, 1042a17.

which is the essence 'of'$_{2\text{-}1}$ Man, and the sense in which Socrates 'has'$_{3\text{-}2}$ an essence, which is the essence 'of'$_{2\text{-}3}$ Socrates. The relation 'has'$_{3\text{-}2}$ may also be expressed as 'exemplifies,' 'is a specimen of,' or (as a relation between the individual and the kind) 'comes in the specific kind ....'

It is the plausible doctrine of Z 6 that the relation 'essence of'$_{2\text{-}1}$ between essence and kind *is* the relation of identity, and that the relation 'essence of'$_{2\text{-}3}$ between essence and composite individual is *not*. (The second conjunct is not stated explicitly, but I infer it to be the line intended on the question 'whether Socrates and essence of Socrates are the same' [1032a8]. Namely, No, Of Course Not. For the essence 'of'$_{2\text{-}3}$ Socrates is, of course, the essence 'of'$_{2\text{-}1}$ Man. That essence certainly is determinative of Socrates in highly important ways, but it and Socrates cannot possibly be the same thing.)

(The very problem of Z 6, 'What things are identical with *their* essences?', in part feeds on the ambiguity of 'their,' which comes from the ambiguity of 'have' in 'the essences they *have*.')

III Form, Matter, and Composite

Short-circuiting much difficult and only intermittently successful struggle with substance-of in *Metaphysics* Z-H, we leap ahead to the eventual result, that substance-of is *form*:

> "what is sought is the cause (and this is the form) of the matter's being some definite thing (τὶ); and this [= cause = form] is the substance" *of* the thing (Z 17, 1041b7-9).

The word for form is εἶδος, which is also translated 'species,' and is also again one of Plato's words for form in *his* sense. The importation of form, and its correlative matter, complicates the picture considerably, for with it the number of entities in the field has gone from two to three and the number of titles from three to five; and it is easy to get the various relationships between the entities mixed up (this is unhappily frequent in the literature). It is especially important to distinguish the relation between the form and the *individual* (the relation 'of'$_{2\text{-}3}$) from the relation between the form and the *matter*. Let us agree on some canonical modes of expression as aids to this end.

(a) *Form and matter*: For the form to the matter, we may preempt the word 'informs,' and agree to say uniformly that the shape *informs* the bronze, and thereby it is that there is a statue of Socrates. (The shape here is 'the cause of the bronze's being some definite thing [τι], viz., a statue; and this is the substance' of the statue. Elegant variations include 'shapes up,' 'overlies as form to matter,' and in the best and most interesting cases, 'ensouls,' or 'empsychs,' or 'animates.') Conversely, for the relation of the matter to the form, the bronze *is informed by* the shape (also 'shaped up by,' 'takes,' 'is subject to [as form],' 'underlies [as matter]'). From here on I will drop 'has,' which once matter is involved is too ambiguous as between a relation of matter to form and the entirely distinct relation 'has'$_{3-2}$ of individual to form.

(b) *Form and individual*: For the form to the individual, let us employ the expression 'is form of,' *never* 'informs' (thus 'Man informs Socrates' is henceforth ill-formed), and agree to say uniformly that the shape *is form of* the statue (not of the bronze). (An acceptable variant is 'constitutes.' At a venture, the *verb* 'to form' belongs here: the shape *informs* the bronze, as just stipulated, and thereby *forms* the statue, as long as we are sure we can say this without conflating the F/I with the F/M relation thereby.) Conversely, the relation of individual to form is as fixed earlier: as its *exemplifying* the form (or 'being a specimen of,' or 'coming in the specific kind ...'). 'Has' is, as just suggested, from now on better avoided. 'Is form of,' of course, coincides precisely with the substantial cases of what in the *Categories* is called 'is said of' (on the proviso that 'is form of' may be used for part as well as all of the form).

(c) *Matter and individual*: This can get complicated, in cases where numerically one and the same individual can survive *changes* of component matter, which are the important cases; and there are related complications to be dealt with later (sec. VI) connected with the division within 'matter' of 'sensible' and 'intelligible'; but we may ignore all that for the moment. Let us say that the matter *composes* the individual (variants include 'is shaped up into,' 'is formed into,' et al.). Conversely, the relation of the individual to the matter is that of *being composed of* ('consisting of,' 'being made of'). Aristotle notes (Z 7, Θ 7) as an adaptation of Greek usage that when an individual is composed of a matter M, the individual may be named *M-en*: thus the statue composed of bronze is braz-en, the box made out of wood is wooden, etc.; this last is no mere terminological nicety but an important piece of metaphysical theory, but it lies off our main track at this point and I will not pursue it here.

It is evident from the exposition this far, however, that I am proceeding here on the basis of a view about Aristotelian form that some would wish to challenge. I am taking it (1) that what Aristotle normally means by form is the specific form that each composite shares with the other members of its species (the ὁμοειδῆ, indeed), and that he does not normally mean an individual form of the composite's very own. This is in fact a very complicated affair, and among other things intersects with one of the most critical unresolved issues in the *Generation of Animals*; I certainly do not wish to wax dogmatic about it; but it must be discussed elsewhere. In addition, I am taking it (2) that the form is a being or ὄν that is to be distinguished from the composite that it is form of, i.e., I think it is a serious mistake in Aristotelian metaphysics to identify the form with the formed thing (and likewise to identify the lack, στέρησις, with the unformed thing), as is done in some recent discussions. I take seriously Aristotle's claim that form is a *cause* of the being that is the formed thing, and that it must always *pre-exist* that which it forms, and that form itself does not come to be. If this be heresy, or alternatively schism, at least I am being explicit about it.

IV The 'This'-Hood of the Substantial Individuals

Aristotle holds that each substantial individual is what he calls 'some this' (τόδε τι). This is a complex and delicate affair; in the interest of ever getting on, I will again have to be more dogmatic than is ideal, and bestow upon you the precipitate of extended reflection on the topic without rehearsing all of the evidence on all of its many sides. What I think Aristotle means by 'some this' in this connection is the following, two-fold phenomenon. First, a substantial individual belonging to a substantial kind F (think of F as Man or Dolphin or Crane) is a well-demarcated and well-worked-up unity of a kind that Aristotle contrasts with what he calls a 'heap' (cf. Z 16 init.); it is strongly characterized by the type of structure or organization called ἀν-ομοιο-μεριστόν, 'composed of unlike parts,' and accordingly there is in the great majority of cases a strong criterion for what it is that counts as *one* F. This phenomenon about substance may for convenience be called its *synchronic individuation*. Second, a substantial individual F's *being F* is necessary or essential to it, in the sense that its coming-to-be is the coming-to-be of an F; its ceasing-to-be-F is its ceasing-to-be altogether, and its persistence as one and the same (to the extent that it does persist) is its persisting as one and the same F — whatever in detail that may consist in, or our criteria for its satisfaction may be. This phenomenon about substance may for convenience be called its *transtemporal* or *diachronic individuation*. (Both of these phenomena are prefigured in the *Categories*, at 3b10-21 and 4a10-b19.)

These two conditions may be thought of as pervasive *explananda*, in this sense: Aristotle's reflections on the problem 'What *is* substance?'

can be seen as efforts to bring to light *what underlying metaphysical frame-work or structure in the nature of things could plausibly be thought to even-tuate in the two phenomena, those of (i) individuality or 'this'-hood, and of (ii) a certain type of natural persistence as the same F through change, that the conditions identify*.

That, in a nutshell, will have to suffice for 'substance means "some this," as we maintain.'

V Form as the Cause of Unity in the Composite

It is well that we have just reminded ourselves of a very fundamental point about the Aristotelian universe, which is that the occurrence in the natural world of these 'thisses', these sharply demarcated, highly organized, integral structures or systems (συστήματα, συστάσεις) — the biological objects which are the substantial individuals, each one a uni-tary individual entity or 'this,' each one exemplifying over its temporal span a sharply defined complete specific nature or essence — is itself a *very remarkable fact* that calls for scientific/philosophical explanation. (Someone might recognize that I am quoting here from myself, for which I apologize, but I don't know how else to say this.[5]) This is so especially because Aristotle adopts an Empedoclean concept of the basic material elements of the world — as opposed to, say, an atomic the-ory — according to which the basic elements are stuffs, bulks, fluids, gases ('Earth,' 'Air'), whose 'combination' is therefore not structural but chemical in character, indeed likened by Empedocles himself (fr.23) to the squeezing together of pigments by a painter. Such elements can mingle and merge, but Aristotle holds that of themselves, καθ' αὐτάς, their nature is not to build up into complex structures and superstruc-tures such as are seen as an autoptic fact in the biosphere; and he is severely critical of Empedocles' story of evolutionary zoogenesis via the commingling of the elements on exactly this ground: the mistak-ing of mixture for structure, the failure to see that *form*, in Aristotle's sense, is required.

Thus far I have tried to keep this discussion of Aristotelian form, matter and individual within the bounds of fairly widespread agree-

---

5 'Transtemporal Stability in Aristotelian Substances,' *Journal of Philosophy* **75** (1978) 624-46; reprinted in D.L. Boyer, P. Grim and J.T. Sanders, eds. *The Philosopher's Annual* (Totowa, NJ: Rowman & Littlefield 1979). Hereafter cited as 'TSAS,' as here, 'TSAS' 633-4.

ment; now for something more avowedly heretical. I believe that there is a deep underlying connection for Aristotle between the notion of being *formed* and that of being *unified* or *one*. In his view, I think, it is form that is responsible, not only for each substantial individual's having[3-2] its permanent essential nature (which is not controversial), but also for its *being* a unitary τόδε τι or 'this,' at all, as opposed to an accretion or aggregation or 'heap' (of this not everyone is yet persuaded). The evidence that inclines me toward this strange-seeming attribution is basically of three kinds, and is what I mainly wish to discuss from this point onward.

First are the many passages, too many to recount here but familiar and long puzzling to all students of the affair, alluding to *form* as τόδε τι, even though Aristotle is in other places well aware that this cannot be right as it stands (possibly the most lucid of these is *Metaphysics* Z 8 1033b19-26). This is a crux of the first rank. I think its resolution is that form is *cause* of the τόδε τι. Here are two fairly clear indications along this line: at *Metaph* Z 17 1041b11 ff., form is responsible for the *syllable*'s being not only whatever syllable it is, but being 'one' as opposed to a heap. And at *de An* II.1 412a6-8, it is form 'in virtue of whch' a substance is spoken of as τόδε τι.

A second indication that *being formed* and *being one* are closely linked or even fused in Aristotle's thought comes from the account of embryonic development that is put forward in the *Generation of Animals*. [I have discussed this in a paper on Aristotelian substances published a few years back;[6] but since it is unlikely that anyone but specialists would have read or remember that, and since I want to say something later (sec. VII) about the sexist aspect of his theory of generation, I will briefly rehearse the matter here.]

Here, as I interpret it, it seems that the substantial differentiae, constituting a newly formed animal offspring as a full-fledged specimen of a substantial species, are also responsible for its character of being a full-fledged individual 'this' as distinguished from an accretion or aggregate or 'heap.' To explain briefly: according to the theory of *GA*, the specific form of the species is stored, in a very special way, in the semen (σπέρμα) of the male parent. The female parent contributes a material mass of quasi-menstrual residue called καταμήνια and also (in most cases) the place where the fabrication of offspring is to occur. The generation of animals is then a process by which the form as stored

---

6 That is, 'TSAS' 636-8.

in the sperma is read into the matter which is the *catamenia*, by way of a pre-programmed sequence of chemical and physical operations, which are discussed in considerable detail but which do not concern us here. In the course of development, the offspring-to-be passes from the condition of the relatively inchoate catamenial mass, through a series of intermediate phases as progressive differentiation takes place under the formative influence of the male contribution, until the final phase is reached, that of the 'completion' (τέλος): one or more fully specified miniature individuals, almost invariably co-specific with the parents (the few exceptions to this, though of some theoretical significance, also do not concern us here). At the earliest stage the embryo is living, but it is not yet actually, though it is potentially, an animal. Subsequently, as the parts articulate out, it becomes recognizably animal, but it is not yet actually, though it is potentially, any specific type of animal. As differentiation proceeds, it moves from more generic to more specific phases; fully specific form is acquired last, and only then, not before, does there actually exist *a man* or *a horse*:

> For it's not simultaneously that *an animal* and *a man* come to be, nor *an animal* and *a horse*, and likewise in the case of the other types of animal; no: the completion (τέλος) comes to be *last* of all, and the most distinctive character (ἴδιον) of each thing comes at the completion of its genesis. (*GA* II 3 736b2-5)

(Depending on the species, the attainment of this fully differentiated, τελεῖον state may be either before or after the time of parturition or its analogue; across the animal kingdom the relationship varies extremely.)

The relevance of all this to my present concern is as follows: suppose we are dealing with the generation of a horse; then the transition described, from the amorphous initial mixture of equine sperma and catamenia to the end product, *the* (new) *individual horse*, has two aspects to it that we today can distinguish from each other, namely (1) the transformation of something undifferentiated and lacking specific form, into a fully differentiated and sharply defined specimen of a substantial species, and (2) the transformation of a 'uniform' (ὁμοιομερές) mass or agglomeration, something of the nature of a 'heap,' into a nonuniform (ἀν-ομοιομερές) entity having the higher order of unity, in which the whole is something beyond the parts, which Aristotle associates with an individual or 'this' (τόδε τι). Now, although these two aspects are conceptually separable for us, as (1) a process that takes matter into specific form and (2) a process that takes catamenial mass into discrete individual, I believe it is useful and important for understanding Aristotle's way of thinking about individuals (not only qua animals but — especially — qua substances) to proceed on the supposition that for him these two aspects are so *merged* into each other as to be distinguishable with great difficulty and frequently not at all. Thus it seems that just as, for him, before full differentiation the developing equine embryo is not as yet *a horse,* but something of a lesser degree of differentiation (though potentially a horse, and destined to become such actually if Nature's plan is fully carried out), so also *it is not as yet an Aristotelian individual 'this,'* but something intermediate between 'heap' and 'this' (though potentially an individual 'this,' and destined to become such actually when differentiation is complete). On this view, the advent of individuality is *pari passu* with that of form, in being gradual rather than instantaneous, in coming by degrees; and one is complete only when, and because, the other is. It is a view on which something's being an individual or a 'unity' in the requisite sense is a phenomenon to be explained, as coming about by something's being 'unified,' which is not to be distinguished — or at best with difficulty — from something's being 'formed'; the degree to which a kind of thing is 'formed' and the degree to which that kind of thing is 'unified' similarly merge together. The substances, in these terms, then, are *most* formed, and *pari passu* are *most* individuated and least heaplike.

This connection helps to clarify a number of very recalcitrant cruxes, as to the relationship between form and 'this.'

VI The Unity of Form Itself

There is a third type of connection between form and unity, this one highly problematic and hard to be clear about, because the signs are that it is Aristotle himself who has problems and is less than clear. This is the question of the unity or oneness of form itself. Many things come together here.

The oneness that is problematical is not that there is only a single form of Man, one of Dolphin, one of Crane, and so on. On this it seems that Aristotle is quite definite (so I am taking it, at any rate, cf. sec. III fin.): there is only one form of Man, which cannot come-to-be or cease-to-be, because by definition coming-to-be is *out of something,* ἐκ τινος, as matter (thus Z 8); what there are many of, and what come-to-be and cease-to-be, are the hylomorphic composites, and what is responsible for the possibility of their coming-to-be and their ceasing-to-be, and their plurality, is their matter (Z 7, 8, 15, H 1, 4, 5, Θ 8).

No: the problem Aristotle has is that the form is *complex*, for it consists of *many differentiae*, and a recurrent ἀπορία, both genuine and serious, in Z and H is: *how is it that the many differentiae combine into a SINGLE, UNIFIED form, a form that is ONE?* Thus,

1. Z 4, 1030a3-4, 10-17, b7-10: definition or formula of an essence is of *one* thing, *not* something predicated of something *else*, *not* a 'participation,' or 'affliction,' or 'accident,' *nor* 'one by continuity,' like the Iliad or things unified by being 'tied together,' but in some other, much stronger way than that.

2. Z 11, 1037a18: it is clear *that* the definition is one single formula, but *on account of what* (διὰ τί) and *by what* (τίνι) is the form that is defined *one* — since it has parts?

3. Z 12, 1037b10: I mean this ἀπορία: that whose formula we call a definition (that's essence, remember), why in the world is this *one*? Take e.g., Man, i.e., Twofooted Animal — let this be its formula. Why, then, is this one and not many, Animal *and* Twofooted? 1037b24: 'the things in the definition have got to be one.'

4. Z 13: 'Arguments' are given that lead to the ἀπορία (1039a14) that substance is incomposite in the sense of simple, 'so that there cannot even be a formula of any substance.'

5. H 3, 1043b32: Substance is like number: both are divisible, and into indivisibles ('for the formulae aren't infinite'), yet, like number, definition has to have something *by which* (τίνι) it is one, and not like a heap; it must be stated what *makes* it one out of many. Also, substance is one, not (as some allege) by being some sort of unit or point, but each is a 'completedness' (ἐντελέχεια) and a kind of nature.

Evidently there is a problem here that impresses Aristotle as real and pressing and difficult, but at the distance we stand from his conceptual milieu, it requires of us a certain effort of imagination so much as to be able to see what it is. It need not be a problem that seizes *us* as requiring resolution in contemporary terms for us to comprehend it in his. I believe it is this. We have seen that he seems to think of form as something that 'unifies' a matter into a 'unitary' τόδε τι, an individual. But if form is to do this, it must itself possess a kind of 'unity' of its own; as the originative principle, ἀρχή, of internal coherence and connectedness, it must itself be 'one.' To put it another way, the type of form that Aristotle thinks must be supposed operative in nature if the phenomena that concern him are to be explained and understood, cannot be just a scatter of 'properties' or 'characters' or 'afflictions,' but a powerful *integrative agency* that is *cause* of the unity, both synchronic and transtemporal, of the natural, i.e., biological, individuals. The many differentiae must come together to make *one* form that forms the many *ones*. 'The substance *of* that which is one, must be one' (Z 16, 1040b17).

Now, Aristotle has two main models for how entities or 'beings' can attach together to make a unity. One is that of an 'affliction' or a πάθος 'coinciding' with a subject, as pallor attaches to a man; such a unity as (a) pale man is a 'coincidental' unity, traditionally, per accidens.[7] This is paradigmatically too weak a unity for the present application (it is the usual example of a *contrast* to the present application, thus Z 12, 1037b13-18), as is indicated by the disparaging comparisons to things 'tied together' or 'one by continuity.'

The other model is that of a form or actuality informing or actualizing a matter or potentiality. We saw in sec. IV that this is evidently regarded as a powerful unifying agency with respect to the hylomorphic composites; but for this to be applied in explaining how the *form* is 'one,' it seems that in some way the actuality/potentiality or form/matter relationship will have to be re-drawn *within the form or actuality itself*.

This is exactly what he tries to do. The attempted conceptualization utilizes two different devices, whose co-extensiveness is not fully apparent (to me, at least); one is that of 'genus as matter,' the other employs a distinction between 'perceptible' and 'intelligible' matter (ὕλη αἰσθητῆ and νοητῆ). Let us look at the second device first.

If you reflect a little on the general schematism of form-matter analysis, the usefulness of such a distinction quickly becomes evident. Aristotle's own example of this in Z 10 is a spoken or written *syllable* as contrasted with a brazen *circle*: here as so often with his artefactual illustrations, it is easier to grasp the distinction independently of his example than by means of it; so let us come back to that. For a better artefactual analogue, consider social clubs or city-states with constitutions, written or unwritten. The *perceptible matter* of such a quasi- (or perhaps pseudo-) substance is the members or the citizens; and it is to be especially noted that this matter can change over time, as club members are inducted and cancel their membership, or as city inhabitants attain citizen status and are banished or die, without affecting the continued existence of numerically the same club or city — an ex-

---

7 In Z 6, some confusion is evident between the attachment of pallor to Coriscus that produces the coincidental unity, *pale Coriscus*, and the combination of pallor and Man that produces the coincidental pseudo-Kind, *Pale Man*. See *Aristotle's Metaphysics: Books Zeta, Eta, Theta, Iota* (Indianapolis: Hackett 1985), 112 (Z 6, 1031a28 n.), or *SFP*, section 23.

cellent analogy with the metabolic exchange of material nutrients and 'useless residues' that is typical for the real, biological substances. However, certain components of such an entity are specified for its organization by its constitution: there shall be a legislative body (or rules committee) chosen in such-and-such a way – an Assembly, a Council; there shall be magistrates, tax officials, archons, ten annually elected generals, and so on – *offices*, which at any given time are occupied by some or other concrete persons, but which, as offices spelled out as part of the ongoing nature of the entity by its charter, are 'parts' that are part of the form; these are what is meant by the entity's 'intelligible' parts or intelligible matter. There is no doubt in my mind that Aristotle thinks of biological form in an analogous way: the composite animal is the form that has 'taken up' the matter (συνειλήμμενον); his phrase, 'by "substance without matter," I mean the Essence' (Z 7, 1032b14), means the form with the matter drained away, the single intelligible form of the species studied by the zoologist, down to the least of its intelligible parts. (Of course, in *nature* the form is found only as form of actual specimens; but the scientist who studies the specimens is interested in the 'substance without the matter.' This is why it is true that 'the ultimate species are the substances [sc. with which we deal], the individuals, e.g. Socrates and Coriscus, are ἀδιάφορα with respect to species,' *PA*, init., 644a23-5). As regards the 'unity,' the idea apparently is that as the perceptible matter is differentiated into a sensible unity *by* the form, so the intelligible matter is organized into an intelligible unity *in* the form. At least, that seems to be how the model works in this case. Whether Aristotle has succeeded in answering his question, '*by what* is the form one – since it has parts?' (Z 11), I am less than certain.

His own example in Z 10, the composite spoken or written syllable and the composite circle of bronze, seems to be thought of in the following way. The *circle* divides into segments, and the segments in turn are of bronze; now, 'the segments are parts in the sense of matter [i.e., perceptible matter] on which [the form] supervenes – although they're "nearer" to the form than the bronze is, when roundness is engendered in bronze' (1035a12-14). The *syllable*, on the other hand, divides into phonetic elements, in just the same way (1035b26-8), and the phonemes themselves are materialized as letters in wax, or movements in air (1035a14-17); but in this case the phonetic elements into which the syllable divides, unlike the segments of the circle, are parts of the *form* of the syllable (1034b26-7, 1035a10), i.e. as intelligible rather than perceptible matter. Schematically, the contrast is supposed to be:

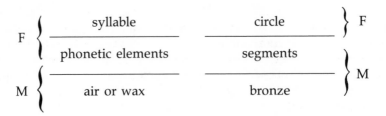

As regards the syllable, the intutition here is marginally apparent: it makes some sort of sense to say that for the syllable-type -LAB-, it is part of the essence or definition that it consist of those phonetic elements -LLL-, -AAA-, -BBB-, in the same sort of way that a certain city-type has to have an Assembly, or a certain animal-type has to have such-and-such a digestive apparatus. As regards the circle, it is less clear why the segment-types of the circle-type have to be sorted with the bronze, and not as something the circle-type has to have (and not just the brazen circle). However, it is bootless to pick further at the example: the principle is tolerably clear and obviously of topmost importance in any case of appreciable differential depth.

The other approach to the problem about Unity of Form utilizes the notion of Genus as Matter. Once again, Aristotle's syllable does not help us much (also, in this case, Z 12, 1038a5-9, it is analyzed rather differently from the other connection). So as before, in order to visualize things better, let us use a different artefactual analogue: automobiles; better yet, American automobiles as they were built in Detroit in the 1950s when I and my contemporaries were auto-manic teenagers. Here the species were: for a start, Buick, Cadillac, Chevrolet, Oldsmobile, Pontiac, all species of the genus General Motors Product; then there were Plymouth, Chrysler, Dodge, belonging to the genus Chrysler Corporation Product; finally the genus Ford Motor Company Product was differentiated into Ford, Lincoln, Mercury; and all of us young American males were adepts at perceiving the three underlying generic affinities, the three basic generic groundplans, that were diversely differentiated into specific varieties in each of these genera. (The μέγιστον γένος, biggest genus, as frequently happens, was ἀνώνυμον, anonymous, without a name, for the three giant manufacturers were just the Big Three. I am leaving aside irrelevant complications like the moneidic genus Studebaker, etc.) We did not know it, but in viewing the automotive kingdom in this way we were profoundly Aristotelian; for Aristotle's concept of genus in its biological application is very like this. Differentia has a vertical as well as a horizontal dimension, having less to do with classification than with construction, in which it bears something close to its modern biological sense, namely, adaptive-in-effect articulation and diversification of structure: it connotes a manner in which relatively undetermined, only generically characterized structures are variously specialized and specified. Genus is an underlying generic potentiality, and the differentiae are the particular manners in which that potentiality is found to be restricted or reduced. (Sorry, I am quoting from myself again, for the same reason as before.[8])

In such a setting, the application to the problem about unity of form is fairly straightforward. As with the 'intelligible matter' device, you start with the composite, and then prescind to the single specific 'substance without the (sensible) matter' (doing the 'matter-drain' again). What you have at that point is the specific form, consisting of a generic component differentiated by a complex variety of differentiae. Its 'unity' is, as Aristotle explains it in *Metaph* H 6, that of actualization (i.e., the differentiae) to potential actualized (i.e., the genus), that is, of actuality to that of which it is the actuality, and it is clear from the *de Anima* that he holds this relationship to be 'the most proper and fundamental of unities' (*de An* II 1). It seems to be important that this potentiality/actuality relation be visualized as *within the specific form itself*; for only so conceived does it resolve the difficulty to which Aristotle repeatedly calls attention in the passages cited (beginning of sec. VI), about what the intra-eidetic structure can be that relates 'the things that are primary.'

In the natural order of things, we should proceed next to see how all this ties together with the theory of biological form designated as ψυχή, or soul. This is complicated and difficult, however, and would take more space than we have left. So I will finish by relating the issues we have been discussing to a difficulty in what might at first sight seem an unrelated area: Aristotle's theory of sex.

VII The Underlying and Interesting Source of the Sexism: A Problem about Unity!

You will recall from sec. V that according to the account of animal generation given in *GA*, the specific form of the offspring is transmitted via the semen of the male, the female contribution being entirely material. This naturally raises the question why it is that the offspring is not

---

8 'TSAS' 635

invariably an identical duplicate of its male parent. Aristotle's response to this is his remarkable theory of Teratology, the causes of deviation from the specific norm. The mechanism of the transmission of form is a pre-programmed set of complex 'motions' in the semen, by which the female's catamenia is 'shaped' and 'set' through the intermediate, generic phases to the completed state of fully specific differentiation (thus the theory is clearly and explicitly *anti-preformationist*; the semen is *form-imparting* without being literally *form-containing* in the sense of containing animalcules). Deviation from the specific norm comes about due to resistance in the passive catamenia to this kinetic activity of the semen; the mildest such deviation is the formation of female as opposed to male offspring, but in cases of more resistant catamenia or of weaker seminal movements, more severe deviations can occur, all the way to the true Terata, the prodigious births and monsters and freaks of archaic legend and rustic imagination. The critically-inclined reader may admire the ingenuity of this, but still wonder how the theory is to explain not mere divergence from sire, but resemblance to dam. Aristotle finally faces this question in GA IV 3, whereupon, suddenly and without warning, the general account is radically altered, and we hear of 'motions' that come from the catamenia (the passage is 767b23-769a6). This is, if not a rout, something of a disorderly retreat; nothing has been heard earlier of 'motions' actively occurring in or coming from the female material in its own right,[9] and the constantly reiterated theme, that it is only the semen of the male that has the 'orgin of motion' (ἀρχὴ κινήσεως) and that the female's contribution is matter alone (ὑλὴν μόνον) has been last repeated only a page earlier (766b12-14). And beyond its inconsistency with the earlier explanation, the GA IV 3 account does not make very good sense in its own right: are the female 'motions' *of* the catamenia (as instrument), or *in* the catamenia (as substrate)? If the former, then *what is being worked on*? If the latter, then *what is working*? Certainly the comparison of 768b15, with hatchets getting blunted or warmers getting cooled, is not enough to support the suggestion of 768a18, that there are female motions adequate to shape the κύημα into a facsimile of mother or mother's mother. It is worthwhile, therefore, to look a little farther into Aristotle's account of the respective roles of the sexes in generation — what factors can they be that have caused him to cling so stubbornly to the male-

---

9 There is not space here for a review of the evidence; the whole subject is comprehensively treated in *SFP*, section 15(iii).

motions-only view, when his own observations (not to say anyone's observations) point so plainly to its untenability that it must be unceremoniously dumped as soon as they are brought up?[10]

The model of the nature of male and female that is taken for a starting-point at the outset of the *GA*, as ἔνδοξον, is that of

male = what has the ἀρχή of movement, and generates in another;
female = what has the ἀρχή of matter, and generates in itself

(I 2, 716a5-7, 13-15); this is their 'difference in λόγος,' from which stem

---

10 Such review of recent literature as I have made, which is anything but comprehensive, indicates that apparently my view of *GA* IV 3 as a collapse of the earlier explanation, though not unshared, is not widespread. Thus D.M. Balme in 'Aristotle's Biology was not Essentialist' (*Archiv für die Geschichte der Philosophie* **62** [1980], 2) equably epitomizes:

> If the sire's movements fail to control the foetal matter, the next most effective determination comes from the female's movements which are present in the foetal matter (blood in the uterus). If they too fail, movements inherited from ancestors exert control in turn ...

The discussion of resemblance in A. Preus, *Science and Philosophy in Aristotle's Biological Works* (Hildesheim 1975), 101-4, evinces no alarm on this issue. P. Pellegrin, *Classification des animaux chez Aristote* (Paris 1982), 134-5, is not concerned with this problem (his focus is upon παρέκβασις as a case of the 'plasticité' of the γένος). J. Morsink, in *Aristotle on the Generation of Animals, A Philosophical Study* (Lanham, MD: University Press of America 1982), calls the introduction of female movements 'an important qualification of the form-matter hypothesis' (138), and argues in its defense against its critics and against sundry defenders of the un-'qualified' theory (141-3):

> The qualifications made in Book IV do not conflict with the initial hypothesis, but bring it into line with some facts that would otherwise be its downfall.

W. Kullmann, *Wissenschaft und Methode* (Berlin/New York: De Gruyter 1974), 52, likewise employs the language of 'modifiziert.' A.L. Peck, in his 'Connate Pneuma' paper (in E.A. Underwood, ed., *Science, Medicine and History: Essays in Honour of Charles Singer*, v. 1, 111-21), seems to think of *GA* IV 3 as an 'extension' and not a contravening of the earlier account. As far as I have found, only Erna Lesky, *Die Zeugungs- und Vererbungslehren der Antiken* (Wiesbaden 1950), 152-4, adequately appreciates the extent of the inconsistency between IV 3 and the earlier treatment. Apart from questions of consistency with the account earlier in *GA*, I have not found company in regarding the IV.3 explanation as wanting in internal coherence (cf. above), nor in tracing Aristotle's persistence to the last ditch in the male-motions-only theory to the theoretical difficulty of dividing the form (cf. below). It seems doubtful that the renegade remarks of *Metaph* Z 7, 1032a28-32, Z 9, 1034b4-6, have much bearing on this issue.

their differences in organic reproductive parts (a23 ff.). When the actual theory of sexual generation is taken up in I 17, the discussion and refutation of the pangenetic hypothesis is the main preoccupation of the next several chapters (17-21); and in the course of this, a suggestion of Empedocles' is considered as to how both the male and female parents might be thought to contribute to generation a genetically significant 'semen':

> Again, if [the semen] comes from all parts of *both* [parents] alike, there will come-to-be two animals; for it will have all the parts of each. Hence, if it's to be stated in this way, it seems Empedocles' view is in closest agreement with this one:[11] for he says that *in the male and in the female there is a sort of tally [i.e., complementary halves in each adding up to a single whole], but the whole doesn't come from either one,*
> "but sundered is limbs' nature, part in man's ..."[12]
> For [otherwise], why don't the females generate out of themselves, that is, if [the semen] comes from the whole body and they have a receptacle? But as it seems, either it doesn't come from all the body, or else it comes in the way that that one [= Empedocles] says — not the same parts from each, and this is why intercourse between the parents is required. (I 18, 722b6-17)

A modern reader can readily see that this suggestion (italicized in the quotation) is from a theoretical standpoint extremely promising: the 'tally' (σύμβολον) is, e.g., one-half of a coin or other conventional item which has been broken in two, one to be kept by each of two parties to an agreement, or one to be carried by a courier to validate, by matching with its complement in the possession of the recipient, the authenticity of his message. But Aristotle does not follow it up, instead criticizing weaknesses in the way it has to be developed within the Empedoclean framework:

> — But this is impossible too. For the parts cannot survive and be alive if "sundered," no more than they can when they are large, in the way that Empedocles generates them in his "time of Love":
> "where many neckless heads sprang up ...,"

---

11 Following Peck, a few words that follow are deleted; but in any case they do not seem to affect the present issue.

12 Presumably this verse of Empedocles' (DK fr. 63, *q.v.*) continued with something like,
'... seed, and part in woman's ...'

> And then, he says, grew together thus [as we see them]. Patently impossible! For not having soul nor some sort of life, they can't possibly "survive," nor, if they are like several living animals, can they grow together so as to become one. Yet this is the sort of thing they have to say who say that [the semen] comes from all the body — as it went then in the earth under Love, so it must go for them in the body. For it is impossible that the parts be produced connected together, and go off together into one place. Then again, how are the top & bottom parts, and right & left, and fore & aft, "sundered"? All these things are non-sensical. (722b17-30, cf. IV 1, 764b15-20)

Thus by the conclusion of the criticism of pangenesis, the initial model of *GA* I 2 is firmly set: the alternative to literal 'parts [of the offspring] in the semen' has been established, namely 'formative motions in the semen' — but the idea of a 'female semen,' which would have 'motions' of its own, is still ruled out. The catamenia, it is explained, is analogous to the semen in every way, but because the female is weaker and cooler, the catamenia is less completely concocted (the discussion is summed up at I 19, 726b30-727a2). The following completes the argument:

> Since this [catamenia] is what is produced in females corresponding to the semen in males, but it is impossible that any creature produce two spermatic secretions at the same time [where, one wonders, did *that* premiss come from?], it is obvious that the female does not contribute any semen to the generation; for if there were a semen there would be no catamenia; but as it is, because catamenia is in fact formed, it follows that semen is not. (727a25-30)

It is in this way, then, that the conceptual bind comes about that leads to the inevitable collapse of this part of the theory in *GA* IV.3: Aristotle has clearly seen some potential merits of the Empedoclean idea that the offspring somehow pre-exists divided between the parents, in the way suggested by the 'tallies' (722b11); but that idea has had to be discarded, as having 'nonsensical' implications (b30). It should be noticed, however, that the implications criticized are *internal to Empedocles's account, which is preformationist*; something of merit obviously remains to the idea that is applicable to the dilemma Aristotle has fallen into.

In fact, the problem he faced was, and is, of major difficulty. Put in the terms of his preferred account of the matter, the problem is: *how the form of the offspring could be (could have been) divided into two (in the parents)*? Or, since according to him the form is imparted via 'motions,' *how the sequence of motions that shape up the offspring once the process gets under way, could be understood to have been directed by a formula ('logos') somehow pieced together from two half-formulas (ἡμίλογοι'?) emanating from the two forebears?* The twofold root of the problem is the exact nature of the halves or 'tallies,' and the manner of their combination into a single unity; and it can be thought of (purely heuristically) as a problem that *Nature* had to contend with long before science began trying to riddle out how Nature had done it. Indeed one way of appreciating its difficulty is to understand something of the way in which it is now thought that Nature solved the problem: by the mechanism of *meiosis* or *reduction division*, in which the form of an animal species (as carried in the chromosomal material) is reduced, for 'spermatic' purposes, by one-half in each parent from a diploid to a haploid condition, and then recombined, diploided, completed, to constitute the genetic coding (the 'formula,' in Aristotelian language) for the offspring.

As a consequence, it seems that the male-chauvinist cast to Aristotle's account of generation may not be wholly, though it undoubtedly is partly, ascribable to unexamined biased assumptions concerning the nature of the sexes.[13] It is true that the model of the sexes assumed as ἔνδοξον is never seriously challenged, and that once the doctrine of concoction of generative residues and the theory of the generative seminal 'motions' are worked out, the corollaries of female 'powerlessness' (ἀδυναμία) (e.g. I 20, 728a18, IV 1, 765b8-15, etc.) and even 'defectiveness' (πήρωσις) (II 3, 737a18, IV 3, 767b8, IV 6, 775a15, etc.) fall out with unfortunate ease. On the other hand, the analysis here points to another source as operational also: that *he was unable to see a way in which the form, i.e., in the context of the general account, the 'motions' conferring form, could be divided.* When, in IV.3, 'maternal motions' make their sudden and entirely unprepared (although, as *we* can see, well-enough motivated) appearance, nothing is said on this — obviously critical — point, and the account suffers from the incoherences already mentioned. But the extreme difficulty of the problem in his theoretical terms goes some way in partially explaining both that deficiency and the overall sexist cast.

---

13 For a thorough lashing of Aristotle on this latter account see M.C. Horowitz, 'Aristotle on Woman,' *Journal of the History of Biology* 9 (1976) 183-213. He is judiciously defended, to the extent possible, by J. Morsink, 'Was Aristotle's Biology Sexist?', *Journal of the History of Biology* 12 (1979) 83-112.

In fact, it has occurred to me that there *was* an answer available, in his terms, that he might have, though he did not, hit upon. There is one 'division into two' in Aristotle's idea of the distribution of form, that could have been theoretically utilized in this desiderated division of form of offspring between *female* and *male forebear*; and that is the division between *genus* and *species or final differentia*.

Here is one way the story could have gone, in which as much of the account as possible is left unchanged. The account of the role of the male remains pretty much as the *GA* gives it. The female contributes a catamenia, also as before. However, the female is further capable of concocting a small amount of her catamenial residue into a formative semen of her own. Because of her cooler nature, as before, however, this female semen is not as powerful as that of the male, and that in two ways. First, male semen is required to start the generative process (this is obviously a point that any theory has to preserve; τὰ θήλεα οὐ γεννᾷ ἐξ αὐτῶν [722b13], 'the females don't generate out of themselves'). Second, female semen lacks the vital heat that is necessary to bring the generation all the way to the ultimate perfection of final differentia; it has a *generic* formative capacity, but falls short of the fully specific. Thus, the male contribution is required to begin the process and to finish it, but between these extremes, ἐν τῇ μέσῃ συστάσει, the two semens are in generally equal combat as together they vie to shape the product. (If grandparental motions, etc., are to be brought in, this is also the place for them.) Because both semens are kinetic and poietic in nature, rather than themselves literally μερη-ἔχοντα or 'part-containing,' the outcome of their collision at each point is *one* 'part,' perhaps resembling one or the other parent; there is no danger of the 'double parts' of the sort that repeatedly loomed on the pangenetic/preformationist basis.

The sole purpose of this little sketch (which it should be emphasized is neither history nor science but sheer fabrication) is to illustrate the point that in the scheme as Aristotle gives it, there *is one* place at which Form *does* divide in two in a way that could be theoretically useful in extricating him from his difficulty: and that is between the 'uppermost' parts of the generic structure or 'chassis' as laid down in the chronological sequence of formations, and the final differentiation via ultimate specific form. We know that he exploits such a relationship in other connections, such as solving the so-called Unity of Definition problem (sec. VI), and explaining the 'most rightful' sense of both *being* and *one*, that involving an actuality's being-predicated of that whose actuality it is (sec. VI b). But all the evidence is that the idea of correlating 'the ἀρχαί of male versus female,' not with 'active versus passive,' but with 'specific versus generic formative activity,' which could have resolved this vexatious quandary in a way basically congenial to the overall view — and when he really needed it — never occurred to him. And so, he had to accept a male-chauvinist consequence, and did so. Like a man.

Summary of Discussion following Furth's paper

[*Pensées de l'escalier* later added by Furth are in brackets.]

ALAN CODE: You distinguish between 'substance' and 'substance *of*,' and then seem to treat the question of what is primary substance as about the former in the *Categories* and about the latter in the *Metaphysics*. But isn't there a single question here?

FURTH: There looks like a single question, but the criteria for being a primary substance are different. In the *Categories*, to be a primary substance is to be an ultimate subject [2b15-17, 2b37-3a1]. That line of approach leads (in the *Categories*) to the substantial individuals as primary substances. But in the *Metaphysics*, the criterion is being that which *makes* something a substance (and thus, being a kind of 'cause') [e.g., Z 17]. That approach leads to the substantial εἴδη as primary substances (and the *Categories* line of approach of course takes you past the substantial individuals to their basic matter [Z 3]). Causal questions are as far as possible evaded in the *Categories*, and of course matter is there ignored altogether.

MARTIN TWEEDALE: I'm not so sure there is this sort of disagreement between the *Categories* and *Metaphysics*. The individuals that you speak of, which are the σύνολα or combinations of matter and form, actually do not concur with the view of the *Categories* that substance does not admit of more or less, i.e., that something cannot be more or less a man, etc. But if you accept that Aristotle allows for individual forms, then indeed you do have something that cannot be more or less what it is, and these seem to be primary substances in both the *Categories* and the *Metaphysics*.

FURTH: I think that when the biology and *Metaphysics* bring in the notion of matter and of the coming-to-be of substances, the theory must drop the claim of the *Categories* that something cannot be more or less a man or whatever. When a living substance comes-to-be, the differentiating κύημα is earlier less man and later more; and the whole theory of τέρατα is built on the idea that the τέρας is less specifically formed than the normal specimen.

TWEEDALE: Introducing the business of coming-to-be reinforces my point, because particular forms are things such that, though they do-not-exist at one time and do-exist at another time, they do not undergo any process of coming-to-be.

FURTH: No. I know people say this, but it cannot be right about Aristotle, or at all. When something is-not and then later is, then it has come-to-be. That is what coming-to-be *is*, what 'coming-to-be' *means*, and that is what the various analyses of coming-to-be try to analyze, such as the Eleatic analysis which says that coming-to-be is impossible because not-being is impossible, or the Aristotelian analysis in terms of the subject, the lack and the form. It is sheer Humpty-Dumptyism to say that something is not and then is, but has not come-to-be, and I don't believe Aristotle means to say it.

JONATHAN BARNES: No. That is not what Aristotle means by 'coming-to-be.' He means that something comes-to-be if there is a time at which you can say 'it is coming-to-be.'

FURTH: If Aristotle means that [that is, tries or wants to mean it — I think it is one of those things that it is impossible to mean], I am deeply alarmed.

[FURTH: The discussion changed direction at this point, but I should try to clarify this part of it a little further. First, in the heat of debate I failed to catch a significant subtlety in Tweedale's last statement, that according to Aristotle some things can not-be and then be, but 'not *undergo any process* of coming-to-be,' and thus misdirected at Tweedale the little tirade that followed (all of which I still maintain). However, I now realize he probably was thinking of the several places where Aristotle says that some things x can *have come to be* (γεγονέναι, in the perfect), without its ever having been the case that x *is coming to be* (γίγνεται, in the present) (thus, e.g., *Meta* E 3, 1043b14-16; a discussion with further citations is in Sorabji, *Time, Creation and the Continum* [Ithaca, NY: Cornell University Press 1983], 11-12. I thank Mohan Matthen for the reference). Now, this is subtle stuff indeed, and I cannot try to pursue here its application to the various 'indivisible' entities, such as points and instants, to which Aristotle applies it (see Sorabji, ibid.). But as far as *form* is concerned, the E 3 passage just cited goes on, 'but it's been proved and made clear elsewhere [reference presumably to Z 8] that no one makes or generates the form, instead a "this" is made, and what is "'out of' these" [components?] comes-to-be,' which has always seemed to me to dismiss the coming-to-be of form or essence out-of-hand (rather than to admit it in the 'subtle' sense). (The topic is also discussed in *SFP*, #20 [iii].) It seems to me that the passages referenced above also rebut Barnes' suggestion as to what 'coming-to-be' *means*, at least as a generalization; but once again I think I also misunderstood him, for he too may not have meant to *assert* to be Aristotle's the doctrine which my 'tirade' was meant to deny (the thing that I think it is impossible to mean).]

JIM HANKINSON: If Aristotle does not allow for individual forms, how is he going to account for the inheritance of traits similar to one's parents? The suggestion you give at the end of your paper, of a way Aristotle could have tried to cope with inheriting traits from mother as well as father, besides seeming altogether bizarre, also does not seem to solve this problem.

FURTH: I agree about the 'bizarre,' and my suggestion was not put forth as having any merit either as an interpretation of Aristotle or as a viable theory of the phenomena. It is just an illustration of how Aristotle's remarks in the *Metaphysics* on the unity of form could have been used to try to handle this biological problem — and more importantly (it may even be defective as an illustration, and the suggestion itself should be junked), it is meant to help people see that the metaphysical problem about unity and the biological problem about divided inheritance are connected. In the *Generation of Animals*, prior to IV 3 there is very little attention to maternal resemblance and no talk of motions in the katamenia.

We do not have to postulate particular *forms* to explain resemblance within a *family*. There is always room in Aristotle's theory for *motions* that are not determined by *form*, in (what I think is) his favored sense of form.

We also should be mindful of the *possibility* of accepting the *possibility* that Aristotle's theory of generation, though raising many of the deepest and most relevant theoretical issues, is a failure as an explanation of the 'causes.'

CODE: I don't see any contradiction between *Gen An* IV 3 and what comes earlier. The difference is not in the theory being given, but in what we are trying to explain.

Also, it is not true that motions in the catamenia are not mentioned earlier than IV 3. The question is whether these motions are *formative*. Well, he sometimes says that some formation does take place just because of the female contribution. The male contribution generates the specific features, but this is not to say that no generative motions come from the female contribution.

FURTH: Generative residues come from blood and blood does have motions in it: as the ἐσχάτη τροφή, that is how it is threptic and auxetic for the various parts it nourishes: that which nourishes Kidney is active and kinetic nephritically, that which nourishes Liver is active and kinetic hepatically, and so on. But these motions are *threptic*, not *poietic*; blood is *not* a semen. Generative residues, including the catamenia, also have motions, but catamenial motions must be threptic too; what is unique to semen (before IV.3) is its having a *poietic* type of kinesis, the power to shape something else. What I think is new in IV.3 is the apparent attribution to catamenia of a type of motion that is co-ordinate with that of the semen; it is this that seems to me both inconsistent with the earlier account and not internally completely coherent, as I stated on p. 92-3. (I suppose it might be asked whether semen is threptically kinetic too; the answer would have to be, No: it is not matter for the animal, it is just form-imparting, and any other motions it might have would be irrelevant for its generative work.)

CODE: I disagree with you on what Aristotle says here. The θερμόν in the semen is transferred and causes the motions in the catamenia. I think that Aristotle in IV.3 is just emphasizing some factors which he had allowed for earlier.

FURTH: Motions in the catamenia that are caused by θερμόν from the male won't do to explain maternal resemblance. To do that, I think Aristotle also has to *unwrite* some of what he said earlier.

The Credibility of Aristotle's Philosophy of Mind

Aristotle's philosophy of mind is at once tantalizingly up-to-date and frustratingly archaic. Some contemporary theorists find it extremely congenial: they see Aristotle as the originator of the functionalist theory of the nature of mental states. Others take a dimmer view of Aristotle's contemporary relevance: they find Aristotle's theory of mind too riddled with outmoded assumptions to be taken seriously any more. Nowhere has this diversity of opinion been more extreme than in the dispute between Myles Burnyeat and the functionalists.

In a provocative paper[1] entitled 'Is an Aristotelian Philosophy of Mind Still Credible?' Burnyeat has brought a powerful line of criticism against those two leaders of the functionalist bandwagon, Hilary Putnam[2] and Martha Nussbaum.[3] Burnyeat is out to discredit not just the Putnam-Nussbaum interpretation but any attempt to read Aristotle's teaching about the soul as an early version of the functionalist theory of mind. If Burnyeat is right, not only Putnam-Nussbaum but also Edwin Hartman,[4] Kathleen Wilkes,[5] and Richard Sorabji[6] are all misguided in their more or less explicitly functionalist interpretations of Aristotle.

---

1 To appear in a forthcoming collection of essays on Aristotle's *de Anima,* edited by Martha Nussbaum and Amelie Rorty.

2 Cf. his 'Philosophy and our Mental Life,' in *Philosophical Papers, II* (Cambridge: Cambridge University Press 1975).

3 Cf. Essay 1 ('Aristotle and Teleological Explanation') in her *Aristotle's De Motu Animalium* (Princeton, NJ: Princeton University Press 1978).

4 Edwin Hartman, *Substance, Body,* and *Soul: Aristotelian Investigations* (Princeton, NJ: Princeton University Press 1977)

5 Kathleen V. Wilkes, *Physicalism* (Atlantic Highlands, NJ: Humanities Press 1978), ch. 7

6 Richard Sorabji, 'Body and Soul in Aristotle,' *Philosophy* **49** (1974) 63-89; reprinted in J. Barnes, M. Schofield, and R. Sorabji, eds., *Articles on Aristotle, Vol. 4: Psychology and Aesthetics* (London: Duckworth 1979) 42-64

But Burnyeat does more than dispute the functionalist interpretation of Aristotle. He also argues that when we correctly understand
Aristotle's philosophy of mind, we will realize that the only thing to
do with it is to junk it. So anyone who finds any contemporary
relevance, functionalist or otherwise, in Aristotle's theory will have to
come to terms with Burnyeat's argument. That is what I propose to
do in this paper. I will try to show that Burnyeat has not succeeded
in refuting either Aristotle or his functionalist interpreters. I will not,
however, attempt to provide additional positive reasons for embrac-
ing the functionalist interpretation.

I will take it that the prima facie basis for the functionalist interpretation is straightforward and familiar. Aristotle's conception of the soul
is biological: soul is that in virtue of which something is alive. A soul,
moreover, is not a Cartesian substance (not a thing in its own right)
but a substance in one Aristotelian sense — the principle of organization of a living body. The soul of a living thing is a set of capacities
in virtue of which that thing lives. In a human, the soul is a complex
set of capacities to nourish oneself, to take in sensory information about
one's environment, to move voluntarily, and to think. It is in terms
of the soul and its actions or movements that we explain these characteristic human activities and account for the bodily parts and systems
on which they depend.

These explanations and accounts are *teleological*. We explain movements in terms of the goals they are aimed at rather than in terms of
the mechanical workings of the body which carries them out. We account for the eye or the heart not in terms of what it is made of but
in terms of its function — what it *does*, what it is *for*.

Aristotle says that the relation of the soul to the body is that of form
to matter. Hence the complex of soul and body — the individual living thing — is subject to the same hylomorphic analysis Aristotle gives
to any physical complex. And a proper understanding of that hylomorphic analysis makes clear how close Aristotle's theory of the soul
comes to contemporary functionalism.

The form/matter distinction is typically explicated by appeal to the
paradigm case of an artifact.[7] A statue is some bronze with a certain

---

7 Relying on the artifact model in explicating the form-matter distinction, as
both Aristotle and most of his commentators do, makes for trouble in understanding his hylomorphic theory of mind. Critics such as Burnyeat and Ackrill
(see below, pp. 118-19) see this as a flaw in the theory; but it might equally
will be taken to be a shortcoming in the model. The problem with the artifact

shape; this house consists of these bricks and boards arranged and assembled in such-and-such a way; an axe is some iron that has the capacity to chop. In the simplest case, form is nothing more than shape; in the more complex cases, form is more like the functional organization of a complex system. But in each case, the substance in question — the statue, house, or axe — is reckoned a compound of matter and form. Bronze, bricks, and iron are matter; shape, arrangement, and capacity are form. In all these cases, matter and form are *contingently* related. In each case, the matter might have had a different form, and the form might have been found in different matter.

Presumably, Aristotle thinks that psychological processes are related to their physiological bases in the same way. *Seeing*, for example, is not just a certain physiological process in the eye; rather, that physiological process is only the *matter* of the psychological process. But if matter and form are contingently related, then there is no *essential* connection between, e.g., *seeing* and any particular type of physiological process. Rather, *seeing* is functionally defined in terms of its object. The object of seeing is the visible, and the visible is color (418a27-9), so seeing is the perception of color. The eye, of course, is noncontingently involved in seeing, and the eye is a physical organ, but that will not be sufficient for us to identify seeing with any particular type of physiological process. For to describe an organ as an *eye* is to describe it *functionally*, as the organ of sight, and does not impose any physiological constraints upon it. The eye of one type of creature may thus be physiologically very different from that of another. Hence it would be *chauvinistic* of us to assume that any creature which sees does so via the same physiological process which serves as the material realization of seeing in us. It is at least possible for vision to have a different material realization in some other kind of creature. This contingent connection between matter and form is what enables Aristotle to adopt functionalism's characteristic anti-chauvinistic stance.

So the key elements of functionalism are clearly present in Aristotle's account. Definitions of psychical states are always to be given in

---

model is that it oversimplifies hylomorphism and ultimately misrepresents it in the cases that are most important to Aristotle. The crucial point of misrepresentation is the contingent connection between matter and form. In all but the simplest cases, matter already contains a great deal of form, and form carries with it many material requirements. (I am grateful to Montgomery Furth for his illuminating presentation of this point during the discussion at Edmonton.)

terms of form and function, never in terms of material composition. Psychical states require some material embodiment, but not any particular kind of embodiment. But there is no unbridgeable gulf between the psychical and the physiological. Soul and body are not opposed kinds of substance with incompatible properties, but formal and material aspects of one and the same thing. Aristotle thus resists hard-nosed materialism but shows no inclination toward dualism. His view seems to be what Bernard Williams has dubbed 'a polite form of materialism.'[8]

Of course it has become a commonplace to point out that Aristotle did not really have a philosophy of mind − did not, that is to say, attempt to solve the mind-body problem as drawn up by Descartes. And since functionalism is a response to the Cartesian problem, there is a clear, but I think superficial, sense in which Aristotle could not have had a functionalist theory of mind.

Burnyeat's criticism of the functionalist interpretation does, indeed, make a great deal of the pre-Cartesian character of Aristotle's thought, but in a striking and sophisticated way. He claims that in recent years it is the mental half of Cartesian dualism that has come under fire, while the matter half has remained intact in all of us. We have inherited our contemporary view of the physical from Descartes. But Aristotle's physics is so different from Descartes' (and, hence, from ours) that no modern philosopher could share it. For in Aristotle's view of nature, the emergence of mind and life from inanimate matter is not something that requires explanation. It is thus Aristotle's physics that is responsible for the fact that his philosophy of mind is no longer credible. New functionalist minds, Burnyeat concludes, do not fit into old Aristotelian bodies.

Now for the details of Burnyeat's argument. He is willing to grant that Aristotle's theory of mind has a superficially functionalist appearance. Aristotle's hylomorphic definitions make it appear that whereas ψυχή and psychical functions must be embodied, there is no special requirement on how they are embodied, or in what kind of matter they are embodied. But this is just the functionalist's point: the mental does not depend on any *particular* material set-up.

Burnyeat focuses on Aristotle's theory of perception, and in particular on the mysterious Aristotelian doctrine that perception involves the sense-organ's taking on the sensible form of the perceived object

---

8 Bernard Williams, 'Hylomorphism,' *Oxford Studies in Ancient Philosophy* **4** (1986), 195

without its matter. A standard interpretation of this doctrine has been advanced by Richard Sorabji:[9] in perceiving a tomato, for example, a portion of the sense-organ, the eye-jelly, goes red. In general, when one perceives a sensible object to be *F*, some part of one's sensory apparatus literally becomes *F*.

The attraction of the functionalist interpretation is that it permits one to reject this piece of antiquated physiology without abandoning an Aristotelian theory of perception. For Aristotle does not *identify* seeing red with the reddening of the eye-jelly. Rather, according to the Sorabji interpretation, the reddening of the eye-jelly is only the material side of perception, the *matter* of which the perception of red is constituted. As such, it stands to the perceptual process of seeing red as does clay to the vase which it constitutes. On this line, Aristotle's account of perception, taken as a philosophical theory, does not depend in a crucial way on his story about the physiological basis of perception. We can thus discard the quaint theory of the reddening eye-jelly, replace it with a more up-to-date physiology, and still claim to be advancing an Aristotelian theory of perception.

Against the Sorabji interpretation of Aristotle's notion of the sense-organ's taking on form, Burnyeat offers a rival account deriving from Philoponus, Aquinas, and Brentano. According to the rival account, the sense-organ's taking on a sensible form is nothing more nor less than an awareness of that form: the eye's taking on a color is just one's becoming aware of a color. Taking *on* a form is to be thought of as taking *in* that form; the sense-organ's becoming *F* is to be thought of as the sense-faculty's becoming *aware* of *F*-ness.

Burnyeat defends the rival account in detail and with great ingenuity. He concludes not only that Aristotle is not telling the particular physiological tale about the material basis of perception that the Sorabji interpretation supposes, but that on Aristotle's account of perceptual awareness, no physiological change is needed for the sense-organ to become aware of a perceptual object. It is this last point that clinches his case against the functionalist interpretation, Burnyeat thinks. For Aristotle would have to be seen as conceding that an animal's perceptual capacities are fundamental, not supervenient. They simply *are* the way they are, and do not require explanation in physiological terms. Aristotle does not regard the emergence of life as a mysterious fact standing in need of explanation. Rather, according to Burnyeat,

---

9 'Body and Soul in Aristotle,' 49: see esp. n.22.

Aristotle has the explanations going the other way around: we explain the physical properties of animals in terms of their contribution to the existence of animal life.

No doubt the proponent of the Putnam-Nussbaum interpretation will be inclined to object that Burnyeat has wrongly saddled the functionalist with precisely the kind of reductionism that he opposes. Far from trying (vainly) to reduce psychological phenomena to physiological (and ultimately chemical and physical) ones, the functionalist agrees with Aristotle that seeing cannot be reduced to any physiological process. But I think that Burnyeat's point can be put in a way that avoids any appearance of confusion of functionalism with reductionism.

What the reductionist maintains, and what both Aristotle and contemporary functionalists deny, is that for any psychological process, ψ, there are conditions, specifiable in purely physical terms, that are necessary and sufficient for the occurrence of ψ. What the contemporary functionalist is especially keen to deny is that there could be physiologically *necessary* conditions. This is what allows different kinds of creatures with radically different physiologial make-ups still to be in the same mental state, a possibility no chauvinistic physicalist with a type-type identity theory can countenance.

However, although the functionalist may not believe that there is any particular kind of physical state that is necessary for a given mental state, he may still hold that the mental state must be 'realized' in some physical state or other. Now not all functionalists would agree with this, but some surely would. After all, some functionalists claim to be advocating a version of physicalism, while others take their view to be incompatible with physicalism.[10] And the physicalistically-inclined functionalist would surely want to say that each token of a functionally-defined psychological state is still a physical state, even if no purely physical description of it will reveal its psychological nature.

Moreover, the functionalist is free, consistent with his anti-chauvinism, to allow that there could be, for any psychological state, *sufficient* conditions, specifiable in purely physical terms, for its occurrence. There might thus be a brain state in humans that is sufficient for their being in pain, and another kind of physical state altogether

---

10 Cf. Ned Block, 'What is Functionalism?' in N. Block and J.A. Fodor, eds., *Readings in Philosophy of Psychology, Vol. 1* (Cambridge, MA: MIT Press 1980) 171-84.

in Martians that is sufficient for their being in pain. But if Aristotle removes physical constraints from psychological processes to the extent Burnyeat says he does, he would be unlikely to follow the functionalist down this path. For the intuition behind the view is that organisms with radically different physiologies may still be in the same (functionally defined) mental state. And the non-dualistic ontology that best fits this intuition is one which denies type-identity but accepts token-identity between the mental and the physical. But Burnyeat's Aristotle cannot accept even token-identity: I may see red even though there is *no* corresponding physiological change in my sensory apparatus.

Burnyeat offers several considerations in favor of his antifunctionalist interpretation of Aristotle's account of perception. At the beginning of *de An* II 12, in characterizing perception in general as the power to receive the sensible form of things without the matter, Aristotle illustrates his point with the analogy of a piece of wax taking on the impression of a signet-ring. But this is the very model Plato introduces in the *Theaetetus* to mark off judgment from perception, and Burnyeat sees Aristotle's use of it as deliberate. In applying Plato's wax block model directly to perception, Aristotle is insisting, against Plato, that perception is something that already includes articulate awareness from the start.

Since perception is awareness from the start, we do not have the problem of explaining how the awareness in perception supervenes on an underlying physiological process. The supervenience of the mental on the physical — the idea that in any two worlds where the physical facts are the same, the mental facts are the same — is a modern invention, and is alien to Aristotle. The underlying physiological processes are merely necessary, and never sufficient conditions for the psychological states they underlie. The boiling of the blood around the heart is a necessary condition for anger, but can occur without one's really being angry, Aristotle notes (403a21-2). Burnyeat's Aristotle, like Plato's Socrates in the *Phaedo*, sees physiological processes as necessary conditions only. Indeed, Burnyeat even thinks that the only physiological conditions relevant to perception, on Aristotle's account, are states of receptivity to sensible form, such as having transparent eye-jelly.

This brings Burnyeat into direct conflict with Sorabji, on whose view it is an entirely physiological process, such as the coloration of the eye-jelly, that Aristotle describes as a sense-organ's receiving sensible form without matter. In so describing it, Sorabji suggests, Aristotle means to be contrasting his own view with that of Empedocles or Democritus, who thought that in vision material particles emanated from the object seen and into the eye of the beholder. Sorabji's Aristotle is thus contrasting two different physical processes, not a physical and a nonphysical one. He describes the reception of form as being *without matter*, but he does not mean that the process is *immaterial*.

Against this account, Burnyeat argues that receiving form with matter is not correctly construed as absorbing some matter carrying a certain form. If it were, then receiving form without matter would be absorbing the form without its being carried by a material vehicle. But this is an absurd way to view the relation between form and matter. Form is not something that can leave one material vehicle (or exist without a material vehicle at all) and be absorbed by another material vehicle. Rather, to receive the form of something is just to become like it in form. Therefore, to receive the form of something without its matter is to become like it in form without becoming like it in matter; and to receive the form of something with its matter is to become like it in both form and matter.

From this understanding of the notions of 'taking on form' and 'taking on matter,' Burnyeat concludes that to receive the warmth of a warm thing, for example, is to become warm without really becoming warm, i.e., to register, notice, or perceive the warmth without actually becoming warm. His idea is this: when you are heated by proximity to a warm object, say, a stove, you become *like* it in both form and matter. Your matter takes on the same form (viz., heat) that the iron of the stove already has. But when you merely perceive the warm stove without being heated by it, you do not become like it in matter, for your flesh does not become like the iron, i.e., does not take on form, heat. Rather, you become like the warm object in form only. You take in its warmth without becoming warm. Or, as Burnyeat is equally happy to put the point, you become warm without *really* becoming warm.

This is a striking interpretation, but Burnyeat does not attempt to offer evidence in its support. Rather, he shows us how he can finesse the single text that would seem to be an insuperable stumbling block. The passage occurs at the end of the crucial chapter *de An* II 12. Aristotle has been discussing the question whether sensible objects, such as colors or smells, can have effects other than being perceived. He seems to be asking what the difference is between, e.g., smelling and the physiological process in which the sensible object, odor, affects the nose. If this is Aristotle's question, then we have him explicitly drawing the distinction between physiological and psychological processes that is crucial to the functionalist interpretation. Burnyeat, on the other hand, takes Aristotle to be asking a different question, viz., what the effect of sensible qualities is on non-sentient things. Aristotle's question, Burnyeat says, is not what more there is to smelling than having an odor affect the nose, but what more there is to odor's effect on the nose than there is to its effect on the air.

The only thing Aristotle says that would seem to rule this reading out occurs at 424b17. Aristotle asks what more smelling is than being affected by something, and suggests in response that perhaps, in addition to being affected physiologically, smelling is *also* perceiving, i.e., being aware. At any rate, that is how his suggestion is to be understood if we read in the 'also' (καί) that one editor (Torstrik) added in a textual emendation. Burnyeat, following Kosman,[11] argues convincingly that the corrupt αι in the defective manuscript reading ὀσμᾶσθαι αι αἰσθάνεσθαι is not the remnant of an original καί but simple dittography. The scribe wrote αι αι αι, which is one αι too many. So Aristotle is not saying that perceiving is something over and above a sense-organ's being affected. Rather, he is saying that the effect on the sense-organ consists in nothing more nor less than an awareness of sensible form. Far from confirming the functionalist reading, Burnyeat concludes, this passage provides evidence against it.

The idea that the effect of sensible form on a sense-organ is nothing *less* than a state of awareness has the consequence, Burnyeat notes, that the matter of which sense-organs are composed is *essentially* capable of awareness. For there is, according to Burnyeat's Aristotle, no physiological state of a sense-organ on which a state of awareness can supervene. Sensible form produces awareness in the sense organ directly; there is no intervention or supervention involved.

But what kind of matter is this that is *essentially* capable of awareness? It is certainly nothing like the matter composing a Cartesian body, whose essence is simply to be extended in space, and whose connection to mind and the mental is as tenuous and contingent as a connection can be. It is in terms of this lifeless, insensible Cartesian matter that the mind-body problem is framed. But how can there be a mind-body problem if there is a kind of matter which has awareness built in at the ground level? And how can a theory be considered a version of functionalism if it denies the contingency of the connection between a psychological state and its physical realization?

---

11 L.A. Kosman, 'Perceiving that we Perceive: *On the Soul III, 2,' Philosophical Review* **84** (1975) 499-519

Burnyeat's case against the functionalist interpretation rests upon these two crucial claims about Aristotle's theory of perception:

(A) a sense-organ's taking on a sensible form is an act of awareness rather than a physiological change, and

(B) it is possible for perception to occur without any associated physiological change.

It is not crystal clear what Burnyeat takes to be the relation between (A) and (B). He offers several considerations in favor of both theses, but it is not always clear which of the two in particular he takes his evidence to support. And while he nowhere argues that (B) follows from (A), his easy transition from (A) to (B), and some of his remarks about them, suggest that he may have the following sort of argument in mind: perception is *nothing more nor less* than a sense-organ's reception of sensible form, and the reception of form is not a physiological process. So since there is nothing *more* to perception than the reception of form, it is possible for perception to occur without any corresponding physiological change.

But this is not a convincing line of argument. Perception may be (i.e., be identical to) nothing more than the reception of sensible form, but there may still be more to (i.e., required for) perception than that. If the eye's taking on the sensible form of the object seen is not a physiological process, that only shows that there is an essential part of vision that is not a physiological process, not that there is *no* physiological process that is essential to vision.

So (B) does not follow from (A). And while (B) is certainly incompatible with a token-physicalistic version of functionalism, the situation with (A) is less clear.

One might also complain, as Nussbaum and Putnam have done,[12] about Burnyeat's argument for (A) as well. For he devotes the bulk of his effort to refuting Sorabji's interpretation of the sense-organ's taking on sensible form. Nussbaum and Putnam reply that even if Burnyeat is right in his criticism of Sorabji (which they seem happy to grant), he will only have established that the reception of form is not the *particular* physiological process Sorabji claimed it to be, but not that the reception of form is not a physiological process tout court.

---

12 Martha Nussbaum and Hilary Putnam, 'Changing Aristotle's Mind' (forthcoming in Nussbaum and Rorty).

However, there is no evidence that in talking about the sense-organ's reception of sensible form Aristotle actually did have some *other* physiological process in mind. Functionalists such as Nussbaum and Putnam should not, therefore, be so quick to distance themselves from the Sorabji interpretation.

A better move for the functionalist, it seems to me, is to concede nothing to Burnyeat without first examining the details of his refutation of Sorabji. This, at any rate, will be my strategy in what follows. I will show that Burnyeat's refutation is flawed and that his argument for the rival interpretation is not compelling. I will also try to establish that the passages in which Aristotle uses the enigmatic notion of a sense-organ's taking on sensible form favor the traditional interpretation.

Burnyeat's case against Sorabji depends on the rejection of Sorabji's account of the notion of taking on form without matter. But Burnyeat's argument against this account is defective. He claims that on Sorabji's account receiving form *with* matter is absorbing matter which carries a certain form, and infers from this that receiving form *without* matter must, on Sorabji's account, be absorbing a form which is not carried by any material vehicle. But this gets Sorabji wrong. On his account, receiving form with matter is not merely absorbing the material vehicle that carries some form; it must also involve, in an appropriate sense, absorbing the form, as well. If you eat a tomato you absorb some matter carrying a certain form, but you do not receive the form along with the matter. Rather, this is a case of absorbing matter *without* form, which Sorabji need not construe as absorbing formless matter, whatever that would mean, but as absorbing matter without absorbing its form. Conversely, the Sorabjian account of absorbing form without matter is absorbing form *without absorbing matter*. Burnyeat's observation that form is not the sort of thing that can move about without a material vehicle is correct but irrelevant — the Sorabji interpretation need not deny it. What is at issue is not whether the form initially has a material vehicle, but whether the form is ultimately received by — i.e., comes to characterize — the matter of a recipient.

In arguing against Sorabji's account, Burnyeat makes a point of understanding the notions of taking on form and taking on matter in parallel fashion. But on his own account there is a curious lack of parallel. On the surface, it all seems quite symmetrical: taking on the form of x is construed as becoming *like x* in form; taking on the matter of x as becoming *like x* in matter. But *being like x in form* is taken to mean being aware of x, while *being like x in matter* is taken to mean having matter that is, in the relevent respect, like x's matter. But my matter is, in the relevant respect, like your matter when my matter has taken on the form that your matter already has. Thus, Burnyeat is able to construe the taking on of the matter of a sensible object — a notion that Aristotle does not discuss — as meaning precisely what, on the traditional view, is supposed to be meant by the taking on of the form of a sensible object.

The dialectical advantage of this manoeuver is that it forces a reinterpretation of the notion of taking on form. If taking on the matter of something means becoming like it in form — becoming *really F* — then what can taking on the form of something amount to? It cannot mean becoming like it in form — becoming *really F* — for that is what taking on its matter is supposed to mean. It can only mean, Burnyeat and company say, becoming F without *really* being F. And what does *that* amount to? Not, as one might suppose, being an illusory, fake, or counterfeit F, but being aware of or noticing F-ness.

But what reason is there to suppose that Aristotle *would* mean by 'taking on matter' what Burnyeat says he would? No evidence is offered to oppose the eminently plausible view that the 'taking on of matter' that Aristotle says is *not* involved in perception is the actual ingress into a subject's body of external material, as in fact happens in nutrition and as is alleged by Empedocles and Democritus to happen in perception.

When we turn to the passages in which Aristotle discusses a sense-organ's taking on the sensible form of the object of perception, we find further difficulties for Burnyeat's interpretation. At 424a1, Aristotle makes the point that in perception the sense-organ is potentially such as the object of perception is actually. On the Sorabji interpretation, his point is quite straightforward, for in perception the sense-organ takes on the sensible form of the object: in perceiving the F-ness of something, the sense-organ itself becomes F. And of course the sense-organ cannot become F unless it is (a) already potentially F and (b) not yet actually F. We cannot feel warmth unless our organ of touch is capable of becoming warm; and we cannot feel the warmth of something our organ of touch is already as warm as. At 424a7 Aristotle goes on to say that the organ which will perceive white and black must itself actually be neither white nor black, but potentially both. Again, his point seems quite straightforward: something which is already actually white cannot *become* white. Perception is a process of taking on sensible form, and the sense-organ cannot *take on* a form it has already assumed.

But what is Aristotle's point on the rival interpretation? Why can't the eye-jelly which is going to perceive white be already actually white?

On the rival interpretation, for the eye-jelly to *be* white is just for the patient to be *noticing* whiteness. But why would Aristotle want to say that one who is already noticing whiteness cannot be about to notice whiteness? Whereas Sorabji takes perception to be, at least in part, a genuine process in which the sense-organ undergoes an alteration, Burnyeat understands it to be not a genuine alteration at all. In perception, according to Burnyeat's Aristotle, the sense-organ is merely brought into activity; perception is nothing more than the exercise of a capacity. This means that the simple logical point about genuine changes, viz., that a thing which is already *F* cannot become *F*, is unavailable to Burnyeat. A thing which is already red cannot be about to *turn* red; but one who is already playing tennis may be about to play more tennis.

A crucial passage for the traditional interpretation is 425b22-6, where Aristotle argues that 'what sees' (τὸ ὁρῶν) is itself 'in a way colored.' This remark makes perfectly good sense on the traditional interpretation. Aristotle is discussing the question of how, or whether, we perceive that we perceive. How can we *see* that we see, when all that we can, properly speaking, *see* is the proper object of sight, viz., *color*? Aristotle's answer is that what sees is in a way colored, 'for the sense-organ receives the sensible object without its matter.'

This as-it-were coloration of τὸ ὁρῶν can be interpreted either, à la Sorabji, as the literal coloration of the eye-jelly, or, following Burnyeat and company, as the visual awareness of color. But it is the coloration of what sees that *explains*, Aristotle says, why perception and images (φαντασία) linger on after the object of perception has been removed. The explanation is simple on the traditional line: we look at a tomato, and the eye-jelly goes red. Remove the tomato and the impression of red persists. This is because something really *is* still red, viz., the eye-jelly.

On the rival interpretation, however, Aristotle's explanation is feeble: the reason the impression of red persists is that we have been *aware of something red*. But this is less an explanation than a restatement of the explanandum. The question is why, when one has been perceptually aware of a red tomato, the *impression* of redness persists after the tomato has been removed. Aristotle thinks he has an explanation of this phenomenon, and on the Sorabji interpretation he does, but on the rival interpretation Aristotle's explanation is a non-starter.

A large part of the motivation for the rival interpretation seems to come from a desire to help Aristotle out, for the account of the physiology of perception Aristotle offers, on the traditional view, is, as Jonathan Barnes puts it, 'open to devastatingly obvious empirical refutation.'[13] Be that as it may (and I think that Sorabji has adequately met this objection), proponents of the rival view do not seem correspondingly concerned about the philosophical shortcomings of what Aristotle has to say on their account of the matter. To think that in the perception of a tomato the eye-jelly literally becomes red may be physiologically naive, but it provides Aristotle with a non-vacuous account of phenomena such as after-images. This is an account that Aristotle cannot give on the rival interpretation. If philosophical vacuity is the cost of avoiding empirical refutation, the rival interpretation's gesture cannot be considered an altogether friendly one.

The only truly recalcitrant passage for the Sorabji interpretation now appears to be the *de An* II 12 discussion of the fact that plants do not perceive. Clearly Aristotle is interested in the case of plants because they are apparent counter-examples to his theory of perception. A plant has a soul and it can take on the form of a sensible object — for example, it can get warm. So why, according to Aristotle's theory, does it not perceive warmth? In his answer, Aristotle must make clear that his theory can distinguish between the effect a sensible object has on a sense-organ and its effect on a non-sensitive subject, such as air, or a plant. And this is just the distinction that Burnyeat's account has Aristotle making.

Of course, Sorabji also sees Aristotle as making this distinction, and agrees with Burnyeat that Aristotle's reason for denying that plants perceive is that their taking on of sensible form is *not without matter*. Where they disagree is over the interpretation of this crucial phrase. Sorabji takes Aristotle to be asserting that plants can get warm only by taking on warm matter; Burnyeat takes him to mean that the only way they can take on warmth is in a *material* way, by having their *matter* become warm.

One may be inclined to agree with Burnyeat here, if only because Sorabji attributes to Aristotle such an implausible theory of plant-warming. Surely Aristotle would have noticed that a plant can get warm by just sitting in the sun, without ingesting any material at all? But Sorabji and Burnyeat may both be wrong on this point. Aristotle says that the reason plants do not perceive warmth is that they do not have a mean (424b2); that is, they do not have the right initial temperature, poised between warm and cold, to perceive these two qualities.

---

13 'Aristotle's Concept of Mind,' *Proceedings of the Aristotelian Society* **72** (1971-72) 101-10; reprinted in Barnes, Schofield, and Sorabji, 32-41

Their matter can get warm, but that material change does not constitute the perception of warmth. The reason it does not constitute perception is not that it is only a material change, nor that it is only achieved by taking on external matter, but that it is the wrong *kind* of material change.

Burnyeat concedes that the requirement that the organ of touch be in a mean or intermediate state appears to support Sorabji's interpretation. His counter-proposal is that the intermediate state of the sense-organ is merely an initial condition required for perception to take place, but that Aristotle does not suppose there to be an actual physical change away from the mean—a warming or cooling, for example — in the sense-organ. Rather, the departure from the mean is what Aquinas called a 'spiritual' change, i.e., a becoming aware of warmth or cold. However, this proposal faces the same problem we encountered earlier at 425b21-6. For Aristotle's explanation of our failure to perceive when our sense-organ is not in the right initial state becomes vacuous on Burnyeat's reading: an already warm sense-organ cannot perceive warmth because it cannot become warm, i.e., because it cannot perceive warmth.

Burnyeat is surely right that a plant's inability to perceive warmth is due to the fact that its matter is not *sensitive* to warmth. But Sorabji is right on the larger issue. For it is still a physical difference between a plant's matter and ours that explains its insensitivity. Perceiving warmth is not getting warm in an immaterial way, but having the right kind of matter — the kind that composes a sense-organ — get warm in a straightforwardly material way.

But this talk of the right *kind* of matter, Burnyeat would surely say, smuggles in a notion that is antithetical to functionalism. For the right matter is matter that is *essentially* alive, *essentially* capable of awareness. And matter that is essentially alive cannot be only contingently related to the form — i.e., to the soul — in virtue of which it is alive.

Burnyeat derives the conclusion that animal matter is essentially alive from two sources. One, which we have already examined, lies in the details of the theory of perception; the other is Aristotle's frequently enunciated *homonymy principle*, according to which a body that is not actually alive is a body in name only, i.e., is not really a body at all, just as an eye which cannot see is not really an eye. It is tempting to treat this principle as a mere linguistic ruling — that, for example, it is inappropriate or misleading to use the term 'body' for what is no longer alive — but Burnyeat understands it as a physical thesis that is incompatible with Aristotle's hylomorphic theory of mind. This tension in Aristotle's thought has been brilliantly articulated by John Ackrill,[14] whom Burnyeat cites with approval.

Aristotle's problem, as Ackrill presents it, emerges when he tries to specify the *matter* component of a living body, i.e., of a hylomorphic compound whose form is its soul. On the one hand, the matter of any compound must *potentially* have that form; on the other hand, it must not have it *necessarily*. It might seem that there is no problem: the matter of an animal is its *body*. But this solution is blocked by the homonymy principle; if we try to pick out the matter without the form, the body without the soul that animates it, we must fail, for if what we pick out is not alive, then what we pick out is not a body. The homonymy principle prevents the fulfillment of the contingent specification requirement. As Ackrill puts it:[15]

> The body we are told to pick out as the material "constituent" of the animal depends for its very identity on its being alive, in-formed by ψυχή.

Nor can we retreat to such candidates as *flesh and bones*, or other such bodily parts and organs, for the homonymy principle applies to them, as well. Here is the way Aristotle puts it (GA 734b24):

> there is no such thing as face or flesh without soul in it; it is only homonymously that they will be called face or flesh if the life has gone out of them, just as if they had been made of stone or wood.

But if we descend to the level of the inanimate elements of which living things are ultimately composed — earth, air, fire, and water — we have gone too far. Although they satisfy the contingent specification requirement, since they are what they are independent of composing a living body, they fail in a different way. For the elements are too remote to be the matter of a living hylomorphic compound; they are not even *potentially* alive (cf. *Metaph* Θ 7). Ackrill concludes:[16]

> Until there is a living thing ... there is no "body potentially alive"; and once there is, its body is necessarily actually alive.

---

14 'Aristotle's Definitions of *Psuche*,' *Proceedings of the Aristotelian Society* **73** (1972-73) 119-33; reprinted in Barnes, Schofield, and Sorabji, 65-75

15 'Aristotle's Definitions,' 126

16 'Aristotle's Definitions,' 132

But this temporal language — 'until,' 'once' — distorts the homonymy principle. Ackrill makes it seem as if the point is diachronic and developmental, viz., that new animals do not come into being by having life installed in previously inanimate bodies. But it is not the point of the homonymy principle to rule out such a Frankensteinian account of the generation of life; rather, it is to remind us of the crucial importance of function in the definition of a living creature or an organic system. The question is not whether there is a time before life begins at which what we have on our hands is a non-living body that is potentially alive; it is, rather, whether we can, in the case of a presently living animal, pick out something that now functions in certain characteristic ways although it will eventually cease to do so, which will continue to exist (at least for a while) after this happens, and whose functioning in those ways is definitive of the life and existence of that animal. What the homonymy principle tells us is that what we pick out for this role cannot be the body.

Still, there is something that looks, acts, and functions very much like the body, although it cannot, strictly speaking, *be* the body, since it will continue to exist after death, when the body no longer exists. Nor is this something the corpse, which only *begins* to exist at death. It is to this continuing something (which non-Aristotelians are inclined to call the 'body') that Aristotle needs to refer. Well, then, let him refer to it in some other way — say, as the BODY. The BODY has accidentally those properties the body has essentially, and in virtue of which the animal is alive. When the BODY functions, the body is alive; when the BODY ceases to function, the body, but not the BODY, ceases to exist.

The hylomorphist's appeal to the BODY does not just pay lip-service to the homonymy principle or treat it as a mere linguistic ruling. But it does, as Bernard Williams[17] has pointed out, leave the hylomorphist with a pair of entities on his hands — the body and the BODY — which are the subjects of psychologial and physiological investigation respectively. And so it seems that the hylomorphist has neatly sidestepped the mind-body problem only to be confronted with the perhaps equally

---

17 In 'Hylomorphism.' I am indebted on several points to Williams' insightful discussion of Aristotle's hylomorphic theory; in particular, I have borrowed from him the distinction between the body and the BODY. I should point out, however, that Williams himself is less sanguine than I about the tenability of a hylomorphic theory.

intractable body-BODY problem. So the hylomorphist is by no means out of the woods.

But at least he is safe from Burnyeat's argument. For certainly the BODY is composed of ordinary matter, and there is no reason to think that the matter composing the body is any different. The difference between the body and the BODY, that is to say, need not be a difference in their matter. In short, we have been given no reason to think that the homonymy principle is a physical thesis to the effect that there is a kind of matter whose life and sensitivity are independent of and not explicable in terms of its physical properties. Granted that a sightless eye is not just not *called* an eye any more, it ceases to *be* an eye. But that is not to say that the *only* difference between it and a living, functioning eye is that one can see and the other cannot. There is still room for a physical difference between the two to account for their functional difference.

Burnyeat has the idea that this is ruled out by the homonymy principle, which he sees as entailing an unbridgeable gap between the physiological and the psychological — between the non-living and the living. If this is how Aristotle intended the principle, we should expect to find him restricting its application to living things. Such a restriction would confirm Burnyeat's interpretation of homonymy and strengthen his conclusion that there is a kind of Aristotelian matter whose life and awareness are built in and irreducible to anything physical.

But Aristotle does not restrict the homonymy principle in this way. For one thing, he seems willing to apply it even to artifacts. Thus, at 412b14-15 he says that an axe no longer capable of performing its function 'would not be an axe, except homonymously.'[18] *Meteor* IV 12 reiterates this point (the example is changed to a saw) and extends it even further into the inanimate realm. What we find is a systematic downward applicability of the homonymy principle, and, along with it, a systematically pervasive appeal to functional definitions. For the homonymy principle is now extended to natural bodies well below the threshold of life and conscious-

---

18 The passage, unfortunately, is vexed. Aristotle suggests this analogy: as a living body is to its soul, so is an axe to its capacity to chop. If an axe were a living body, this capacity would be its soul, whose removal would render it no longer an axe, except homonymously. 'But in fact,' Aristotle goes on, 'it is an axe' (νῦν δ'ἔστι πέλεκυς). The commonest reading of the quoted sentence takes it to withdraw the counterfactual assumption: an axe is *not* a living body, so it doesn't have a soul — it's just an axe. But on another reading, it refers back to the consequence derived from that assumption: since an axe is not a living body, it remains an axe even when it can't chop. On the second

ness – viz., all the way down to the elements themselves (390a7-19):

> [E]ach of the elements has an end and is not water or fire in any and
> every condition of itself, just as flesh is not flesh ... What a thing is is
> always determined by its function: a thing really is itself when it can
> perform its function; an eye, for instance, when it can see. When a thing
> cannot do so it is that thing only in name, like a dead eye or one made
> of stone ... So, too, with fire; but its function is perhaps even harder
> to specify by physical inquiry than that of flesh. The parts of plants, and
> inanimate bodies like copper and silver, are in the same case. They all
> are what they are in virtue of a certain power of action or passion –
> just like flesh and sinew.

Aristotle thus insists on functional definitions even of copper and sil-
ver, water and fire. His doctrine concerning inorganic compounds and
their component elements, then, is not in principle different from that
concerning animals and their parts; he draws no sharp distinction between physicalistic and functionalistic descriptions and explanations.
It is sometimes thought that this tends to discredit the functionalist
interpretation, on the grounds that functionalism requires just such
a contrast between functional and (merely) physical explanation that
Aristotle seems unwilling to provide. But this line of thought has no
force unless the version of functionalism envisaged is one that is incompatible with physicalism; and there is no reason to suppose that
Aristotle's theory would be of that variety.

The lesson we learn from *Meteor* IV 12 is instructive on this point.
Aristotle indicates that he regards the matter of which living bodies
are composed as inherently no different from that composing the rest
of nature. His conception of the physical is not so deeply alien as Burnyeat has supposed; in particular, it does not prevent his hylomorphism
from being plausibly construed as a direct ancestor of contemporary
functionalist therories.

---

reading (but not the first), Aristotle refuses to apply the homonymy principle
to the axe. But the first reading is preferable, as becomes clear from Aristotle's
justification: 'for it is not of this kind of body that the essence or formula is
the soul, but of a certain kind of natural body having within itself a source of
movement and rest' (οὐ γὰρ τοιούτου σώματος τὸ τί ἦν εἶναι καὶ ὁ λόγος ἡ ψυχή, ἀλλὰ
φυσικοῦ τοιουδί, ἔχοντος ἀρχὴν κινήσεως καὶ στάσεως ἐν ἑαυτῷ). Cf. R.D. Hicks, *Aristotle: De Anima, with translation, introduction and notes* (Cambridge: 1907), 316-17.
(I wish to thank David Keyt for a helpful discussion of this passage and for
convincing me that the favorable reading is in fact the right one.)

Summary of Discussion following Cohen's paper

MOHAN MATTHEN: I am inclined to think that Aristotle was a functionalist about *some* things, but I have difficulties seeing how functionalism is to be applied to perception, especially at the level of a state like seeing red — that is, at the level of perceiving what Aristotle called a proper object of sense. In order for functionalism to be plausible at this level, one has to maintain that, e.g., sensing redness is significant enough teleologically that if some organism should fail to possess the red-sensitive mechanism we happen to possess, then it would be likely to possess some other red-sensitive mechanism. But this is not very plausible: more likely a creature without our red-sensitive mechanisms would be sensitive to something else, to some other significant property of the environment. So the fact that we sense *redness* is due to the wholly contingent fact that our evolutionary predecessors happened to possess the precursors of cone-cells. The choice of the direct objects of sense are thus much more sensitive to mechanisms than a functionalist can comfortably allow. To bring this down to Aristotle: what evidence is there for thinking that he takes the eye-jelly to be just a sufficient but not a necessary condition for being receptive of colors? Quite likely this is a question that tells against Burnyeat as well: for Aristotle might have made our sensitivity to red more dependent on our physiology than he allows possible.

MARC COHEN: I think this is a very good objection to any non-physicalist account of Aristotle's theory of color perception. Aristotle may think of some mental states as having physiological bases that are sufficient, but not necessary, for those states to obtain, but color perception probably isn't one of them. It's more likely that he thought that the relevant sufficient condition (the coloration of the eye-jelly) was also necessary. But, as you note, this objection also presents a problem for Burnyeat; if the physiological mechanisms of perception are essentially involved in an account of what *redness* itself is, then they will also be involved in the *awareness* of redness, contrary to what Burnyeat supposes.

ROGER SHINER: Your use of the body-BODY distinction suggests that Aristotle thought of humans coming-to-be in much the way that Frankenstein's creature came-to-be: that is, by an inanimate body becoming suddenly animate. I think we should avoid ways of talking that implicate him in this way of thinking.

COHEN: I certainly did not intend the body-BODY distinction to suggest that Aristotle subscribed to a Frankensteinian theory. Bodies are not BODIES, and what was once a body can cease to be one and become (or, perhaps, be replaced by) a BODY. But this is a one-way street: a BODY can never turn into a body. So here I am able to use the body-BODY terminology to express Aristotle's anti-Frankensteinianism, if I may call it that.

I did say that it was not the point of the homonymy principle to rule out the Frankensteinian account. Of course, I did not mean to suggest that Aristotle wants to safeguard such an account. My point was only that this is not what he meant to be ruling out when he deployed the homonymy principle to show that the body would not be a body without the soul.

JIM HANKINSON: What is the relation between functional explanation and the matter that embodies the function? Aren't the functional properties, even those of artifacts, necessarily tied to some sort of matter? And if so, how does the functionalist pull off multirealizability?

COHEN: That's a good question, because, in a slightly different form, it's one which Aristotle actually worries about. For him, the puzzle is whether there should be reference to matter in a definition, and he seems to vacillate on this issue. In *de Anima* (A 1) and *Metaphysics* (Z 7), for example, he is willing to include reference to matter in a definition; in *Metaphysics* (H 4) he observes that some substances have definite material requirements ('a saw could not be made of wool ... or of wood' – 1044a28). But in *Metaphysics* (Z 10-11) he claims that 'only parts of the form are parts of the definition' (1035b33). The most straightforward way to reconcile these apparently conflicting tendencies is to say that matter terms may sometimes occur in definitions, but when they do so they refer to form rather than to matter. The idea here is that what makes a certain kind of matter requisite for a certain kind of substance has to do with the *form* of that matter, which would include its functional characteristics.

I think this tells us how a functionalist can pull off multi-realizability. He must claim that when a functional property (e.g., of an artifact) is necessarily tied to some sort of matter, it is tied to it by virtue of some functional (i.e., formal) characterization of that matter.

Thus, axes are tied to iron not qua iron, but qua hard material capable of holding a sharp edge. Iron was once presumably the only suitable material to make axes out of, so it was then not obviously inappropriate to include *iron* in the definition of *axe*. But *iron* was merely going proxy for a functional charaterization of the required *kind* of matter. (I doubt that Aristotle would reject an axe made of stainless steel as definitionally defective.)

MONTGOMERY FURTH: Here is Aristotle's view of the structure of an animal:

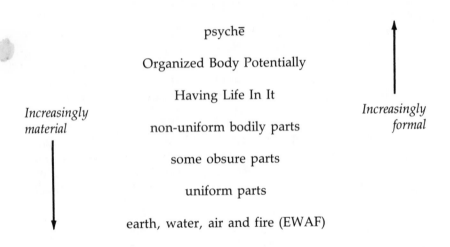

psychē

Organized Body Potentially

Having Life In It

Increasingly material

non-uniform bodily parts

Increasingly formal

some obsure parts

uniform parts

earth, water, air and fire (EWAF)

Ackrill's point is misguided because the soul, i.e., the form, reaches way down into the matter — and most matter disappears when soul does. Talk about a special mysterious life-possessing matter is unnecessary if you think of a form that integrates matter to this extent.

COHEN: I take it that you agree with me that Aristotle has no peculiar kind of matter that separates his physics radically from ours.

FURTH: Burnyeat is trying to avoid the Frankenstein picture when he introduces that matter.

COHEN: But that's just to replace one implausible and mysterious theory with another. Even if any organism capable of perception and cognition must be made of a special 'organic' matter, that matter is itself composed of ordinary inorganic matter, and ultimately of the four elements.

Your diagram does a good job of making that clear. It also helps us to understand why the Frankensteinian theory and the 'special matter' theory do not exhaust the alternatives. My only objection to the diagram is that it suggests that after death all we are left with is the elements. In fact, it seems that what we are left (at least for a while), viz., a corpse, lies at some higher level of organization than that of the elements, even if it is not alive.

FURTH: What we have left is rubble, which we are tempted to call homonymously a man.

ALAN CODE: If functionalism says that the relation between the psychical function and the material physiology is contingent, then Aristotle's view cannot be functionalism. But can't we say that the whole depth of soul or form (in Furth's diagram) is a functional state of the material elements, granting Aristotle does not think that there are even possible alternative material compounds for this functional state. This view would be functionalist in the sense that the soul is a functional state of matter.

COHEN: this is a tempting move to make if we want to break, once and for all, the awkward non-contingent connection between the soul and its material basis. We will then have just one level of material components, viz., that the elements, and something very complex as the formal component. Now, since we have gone all the way down to the bottom (in Furth's diagram) we don't have to worry about a conflict with the homonymy principle. Flesh may be necessarily actually alive, but earth, air, fire, and water are not. However, I doubt that Aristotle would be happy with this conclusion, since in his view the elements are not even potentially alive.

CODE: The elements are not potentially alive only in the sense that they are not the *proximate* matter for living body. In a broader sense we can say they are potentially a living organism. Aristotle is basically looking at the relation of a substantial form to inanimate matter. The ergon comes from the form. So we have a functional organization of an inanimate substratum. Aristotle looks at it this way because his physics is so crude.

FURTH: Θ 7 says that nothing is potentially x unless it is the immediately preceding matter. It is a very valuable passage because it is the one place where Aristotle clearly discusses the vertical dimension of ἐξ οὗ.

Metaphysics and Logic

I What is Metaphysics?

This paper is intended to make progress towards a general, unified characterization of Aristotelian metaphysics.

First, although metaphysics does study the principles that apply to beings qua beings, it also treats of various principles that do not apply to everything. An example would be the principle that it is impossible for contraries to simultaneously belong to the same subject. This is a principle *about* one of the terms that belong to being as such: the contrary.

Second, the metaphysician must *both* state definitions of such terms *and* demonstrate their ἴδια and per se accidents. The principle just stated about contrariety is part of the account of what the contrary is, for contrariety is a type of opposition, and opposites cannot belong to the same thing at the same time. A per se accident of the contrary would be the theorem that each contrary has exactly one contrary. This follows from the definition of contrariety as the *greatest* difference.

Third, the metaphysician must *both* state the general (propositional) principles that apply to being as such *and* treat of their properties or features. An example would be the principle of non-contradiction (PNC). One of its features is that it is the firmest of all principles, another is that it is a prerequisite for rational thought and discourse.

Fourth, metaphysics as the general science of being employs a methodology that first attempts to find indemonstrable, basic applications of various terms or concepts, and then explains the other applications of these terms by relating them to their basic applications. For instance, the metaphysician singles out substance as the primary instance of being, and explains other modes of being by relating them to the primary instance. Likewise, in metaphysics one can draw a distinction between necessities such as 'Socrates is necessarily a man' and those like 'Socrates is necessarily receptive of grammar,' taking the former as basic, and explaining the latter by reference to the former.

1 Beyond his Platonic inheritance

A) *Two general Platonic principles about science, and the need for formal logic.*
In Book A of the *Metaphysics*, Aristotle conceives of metaphysics as a
universal treatment of the causes and principles of things. Such a
science would involve a systematic account of the ways in which causes
and principles are employed within the various branches of knowledge
so as to yield understanding. This general project he inherits directly
from Plato, and in so doing he takes over a number of Platonic assumptions.

First, he takes over the Platonic idea that knowledge (at least in the
strict sense) is knowledge of that which could not possibly be other
than it is, and hence is knowledge of necessary truths. Second, Aristotle
accepts the idea that within the body of a science one may divide the
(necessary) truths into those that are in some sense *not* explained or
'caused' by anything (these are the principles that Aristotle would clas-
sify as 'known through themselves'), and others that are explained or
'caused' by something other than themselves. The thought is that one
can exhaustively partition the truths of some science into those that
are known through themselves, and those that are known through
other things (i.e., things other than themselves) that are their causes.
Knowledge, or understanding, is *systematic* in that the truths that are
not known through themselves are known by tracing them back to
those origins (ἀρχαί) and explanatory factors (αἰτίαι) that are their causes.

So much would be common ground to both Plato and Aristotle.
However, Plato has no *general* account as to how one gets from first
principles to the theorems (as I shall call them). A perfectly general
account of the way in which causes and principles are employed in
such a way as to yield understanding must be able to answer the question: how does a systematic understander get from those truths that
are known through themselves to those propositions that are known
through their causes?

Given the assumption that knowledge is of that which is necessary,
both the first principles and the theorems must be *necessary* truths.
Aristotle thinks that one knows a theorem when one sees that it *follows of necessity* from those principles that are *its* 'causes.' Thus, if one
is to give a general characterization of the way in which causes and
principles are employed in the sciences, a general characterization of
the relation *'follows from of necessity'* is mandatory. Furthermore, reasoning from necessarily true principles to necessarily true conclusions takes
place in a variety of sciences that do not (according to Aristotle) share
a common genus or kind. Hence a perfectly general account of this

relation must in no way depend upon the particular kind of subject matter with which some science deals. The account of the relation must abstract from the particular content of the propositions of any given science, and in this sense the account will be *formal*.

Without playing down the role that the actual practice of dialectical debate played in the invention of Aristotle's theory of the syllogism, we can see that Aristotle had a strong metaphysical/epistemological motivation for inventing a formal logic. It is in the *Prior Analytics* that he presents this system. Having characterized the syllogisms of the first three figures, one of his chief aims is to reduce the so-called 'imperfect' syllogisms of the second and third figure to the 'perfect' syllogisms of the first. The perfect syllogisms of the first figure are supposed to be basic, obvious cases of the relation 'follows from of necessity,' and the imperfect syllogisms are supposed to be genuine, but non-obvious, cases of that same relation.[1] The reduction shows that in principle any reasoning that uses second or third figure resources to derive a conclusion can be replaced by reasoning that derives the same conclusion from the same premises relying upon only obvious inferences.

B) A general strategy in metaphysics. Notice that the reductive strategy of the *Prior Analytics* does not yield two distinct meanings for the word 'necessary' as applied to propositions. The word has exactly the same meaning regardless of whether it is applied to a first principle or a theorem. Nonetheless, a theorem, unlike a first principle, is necessary by virtue of the fact that it follows of necessity from other necessary truths.[2] The word applies primarily and strictly to the first principles, in a manner that requires no explanation, but applies to the theorems in a manner that must be explained in terms of an appropriate relation ('follows from of necessity') to principles.

The same type of strategy is employed in *Metaphysics* Γ 2. *First*, it is used to explain how there can be a unified treatment of being. He shows that despite the fact that there is no single condition in virtue of which all beings are properly called 'beings,' there can nonetheless be a single science of being — a science that studies both the primary beings and those things which owe their existence to them. The word

---

1 See ch.1 of Jonathan Lear, *Aristotle and Logical Theory* (Cambridge: Cambridge University Press 1980).

2 See *Metaph* Δ 5, 1015b9ff.

'being' has a single meaning in its application to substances, qualities, quantities, and so on, and yet it applies primarily to substances, and derivatively to all else. The application of the term 'being' to things other than substances must be explained by relating them in appropriate ways to substances, the primary instances of being. *Second*, this is the strategy that the metaphysician utilizes to investigate all of the per se characteristics of being (such as unity, sameness, otherness, contrariety and the like).[3] Its employment in this connection can be used to show that metaphysics, the general science of being, is also the science that treats of the basic logical principles. It is to this topic that I now turn.

2 Common axioms and the nature of metaphysics

A) A metaphysical puzzle about the study of logical principles. Metaphysics is a universal science (ἐπιστήμη) that investigates the causes and principles of things.[4] When we consult *Metaphysics* B we find that a number of the puzzles, or ἀπορίαι, that structure metaphysical investigation are puzzles about metaphysical inquiry itself. Four of the first five puzzles[5] stated in the first chapter of that book are explicitly about the nature of the metaphysical enterprise, questions *about* metaphysics. Each is a puzzle about the further characterization of that universal science that studies or treats of the causes and principles of things in the most general, or universal, way. The investigation of each of these puzzles takes place within the very discipline that serves as its subject matter.

The second puzzle[6] listed in *Metaphysics* B 1 assumes that this science of causes and principles will at the very least deal with the principles of substance (οὐσία), and then goes on to inquire whether it will *also* deal with the common axioms, those principles 'from which everybody makes proofs.' For instance, should it consider whether it is possible simultaneously to assert and deny one and the same thing?[7] When this

---

3 *Metaph* Γ 2, 1004a25 ff.

4 *Metaph* A 2, 982a21-2 with 982b9-10 (cf. Γ 1, 1003a23-4; Γ 3, 1005a35 without Jaeger's περὶ τὸ)

5 *Metaph* B 1, 995b4-13, 18-25

6 *Metaph* B 1, 995b6-10; cf. B 2, 996b26-997a15

7 Both the principle of non-contradiction and the law of excluded middle are listed as examples of common beliefs at 996b28-30.

puzzle is further elaborated in B 2, Aristotle brings in a third option, suggesting that perhaps there is no science at all that studies the common axioms.

This same puzzle is raised again in *Metaph* Γ 3, 1005a19-21, where Aristotle argues that the science of substance is indeed the science that studies the common axioms (1005a5-8). Here, unlike in B 1 and B 2, he appears quite confident as to the solution of the puzzle. After having explained in Γ 2 that metaphysics investigates being and the per se properties of being (and hence must begin with a study of substance, the primary instance of being, and the per se properties of substance; see 1005a13-14), Aristotle opens Γ 3 with an allusion to B 1, 995b6-10, asserting that we must state whether the science of substance just described also investigates the things that are called axioms in the mathematical sciences (1005a19,21). It does investigate the axioms, he tells us, because the science of substance is (as he thinks he has already established in the previous chapter) the general science of being qua being (1005a13-18), and this latter is the science that studies what belongs or holds good (ὑπάρχειν) per se of all things that are (1003a21-2). Each common axiom holds good of all things that are qua things that are (1005a22-3). These axioms hold good of *all* things, and do not have an application merely in one particular kind (γένος) apart from the rest of what there is (see also a27-8). Thus the study of the common axioms falls within the scope of metaphysics.

We might be tempted to accuse Aristotle of equivocation in his use of the term 'belongs.' It is not obvious that a propositional principle like the PNC *belongs* to a being in the same way that properties *belong* to it, and hence perhaps a science that investigates the various *terms* that belong to all things qua things that are, need not also study those propositions that 'belong' to beings quite generally. We can gain further insight into Aristotle's claim that metaphysics studies the PNC by seeing why that science must study contradiction.

B) The principle of non-contradiction. (i) What a contradiction is. Roughly speaking, the PNC is the principle that it is impossible for the same thing (predicate) to belong and not to belong to the same thing (subject) at the same time and in the same respect — it is impossible for a contradiction to be true. What, then, is a contradiction?

According to the account given in *de Int*6, 17a33-4, a contradiction (ἀντίφασις) is a pair of opposed statements (ἀντικείμενα), one of which is an affirmation (κατάφασις), the other of which is a denial (ἀπόφασις). The affirmation and the denial are statements having the same subject, but what is affirmed of the subject in the affirmation is precisely what is denied of that same subject in the denial.

An ἀντίφασις is not the statement formed by the conjunction of the affirmation with the corresponding denial, rather it is a pair of statements. To believe an ἀντίφασις is not to believe a single conjunctive proposition. It is to have two separate beliefs: one a belief corresponding to the affirmation, the other a belief corresponding to the denial.This point will be relevant to the argument in Γ.3 that it is impossible to believe a contradiction. He is not arguing directly against the claim that it is impossible to have the conjunctive belief, but instead is considering *two* beliefs.

Not only does Aristotle apply the term ἀντίφασις to pairs of statements, he also applies it to pairs of items that may be present in a subject. At *Metaph* I 7, 1057a34-6, we find 'contradiction' characterized as an ἀντίθεσις (another word that is often translated as 'opposition') such that for anything whatsoever, one part or the other of the ἀντίθεσις is present, there being nothing between the two members of the ἀντίθεσις. That is, it is an opposition to which the law of excluded middle (LEM) applies.

Confined to unquantified statements, this duality of usage is quite harmless. Two statements will be contradictories in the first sense just in case the items ascribed to the subject are contradictories in the second.

In *de Int* 7 the notion of a contradiction is extended to cover universal statements. At 17b16-20 we are informed that the universal affirmative and the particular negative statements are contradictories, as are the universal negative and the particular affirmative. In the case of 'universals not taken universally' Aristotle actually allows that both members of a contradiction may be simultaneously true (17b29-37). For instance, '[a] man is pale' and '[a] man is not pale' (construed as statements that are not taken universally, although the subject is a universal) are contradictory statements. The discussion of contradiction in this paper will be limited solely to unquantified statements, and the further complications arising from universals not taken universally will not be pursued here.

(ii) Why does metaphysics study contradiction? (a) What are the per se attributes of being? Having already asserted that there is a single, unified science that treats of being qua being,[8] and having explained that

---

8 *Metaph* Γ 1, 1003a21-2; Γ.2, 1003b15-16; cf. 1005a2-3, 13-14

such a science must study the causes and principles of substance,[9] Aristotle argues in the remainder of Γ 2, 1003b11-1005a18 that the science in question also studies the various ubiquitous terms that apply to being as such — terms such as 'one,' 'many,' 'same,' 'other,' 'similar' and 'dissimilar.'[10]

The opening sentence of Γ 1 considers items belonging per se to being (1003a21-2) as falling within the scope of the science, but it fails to tell us which terms those are. In Γ 2 he argues that 'one,' 'many,' 'same,' 'other,' 'similar,' 'dissimilar,' 'equal,' 'unequal,' 'different' and 'contrary' are such items (and he adds without further argument 'complete,' 'prior,' 'posterior,' 'genus,' 'εἶδος,' 'whole' and 'part'), and hence are to be studied by the general science of substance. Γ 2 calls these the καθ' αὐτὰ πάθη (per se characteristics) of being (qua being) and unity (qua unity) at 1004b5-6, and the ἴδια (properties) of being (qua being) at b15-16.

(There are further subtleties involved in specifying the terms investigated by the science of being. *Metaph* B 1, 995b18-27 first asks a question not explicitly solved in Γ 2: does the science of substance also study the per se accidents of substances [τὰ συμβεβηκότα καθ' αὐτά, b20]? It then goes on to ask an apparently distinct question [πρὸς τούτοις] whether it will study in addition to these such terms as 'same,' 'other,' 'similar,' 'dissimilar,' 'contrariety,' 'prior' and 'posterior.' Finally it ends by asking whether it will also study the per se accidents of these last mentioned items — that is, in addition to investigating the definition [or 'what it is'] of each of them, will it also study such issues as 'does each contrary have a single contrary?'[11])

---

9 *Metaph* Γ 2, 1003b17-19; see also 1004a31-3, 1005a2-3, 13-14. In every case, each science is, strictly speaking (κυρίως), the science of its primary object — this being that upon which its other objects depend, and that by reason of which they are called (what they are called with reference to that science) — 1003b16-17.

  For instance, the science that studies healthy diets, healthy medicines, healthy complexions and healthy bodies is (strictly speaking) the science of *health*, for their being what they are called (namely, 'healthy') consists in their standing in appropriate relations to a single thing: *health*. It would be a mistake to say that the study of these things is something additional to, over and above, the study of health. In the same way, since *substance* is the primary object of the science of being, strictly speaking the science of being just is the science of substance.

10 This is in response to another puzzle from *Metaph* B.1 (995b18-27; see also the related puzzle, unsolved in Γ 2, at B 2, 997a25-34).

11 *Metaph* B 2, 997a25-34, which elaborates upon 995b18-25, discusses only the

(b) How and why does metaphysics study the per se attributes of being?
To study the per se properties of being, the metaphysician must both
state their definitions and prove theorems about them in a manner at
least analogous to geometrical proof. Regarding each such term, he
must know *what it is* (i.e., he must be able to state its definition), and
know its properties (ἴδια) and per se accidents (τὰ συμβεβηκότα καθ' αὑτά).
For instance, in geometry one defines the concept of a triangle (states
what a triangle is; perhaps, 'three-sided polygon'), and next proves
theorems such as the right-angle theorem (that triangles have interior
angles equal to the sum of two right angles). The theorem specifies
a property that necessarily holds good of all and only triangles. The
metaphysician, when studying the ubiquitous terms that are per se
accidents of being, proceeds in an analogous fashion, defining each
term and then demonstrating its per se accidents. For instance, he will
define *sameness* and then prove theorems about it. To this extent, at
least, metaphysics is a science conforming to the *Posterior Analytics*
model.[12]

Although metaphysics does study principles that apply to all things
that are solely in virtue of being things that are, and does not study
any of the principles that hold good of particular kinds of things that
are qua some particular kind of being (1003a21-6), nonetheless it *also*
treats of certain principles that do not apply to absolutely everything.
That is, metaphysics does not study *only* ubiquitous principles that apply to all things. However, definitions of ubiquitous terms, as well as
theorems involving them, will be related in a fairly close manner to
propositions that *are* true of all things. The definition of a non-
ubiquitous term like whiteness will be intimately related to a proposition about the things of which whiteness is an attribute. If whiteness
is a penetrative color, then white things will be penetratively colored.
In much the same way, we should expect principles about ubiquitous
terms to be related to ubiquitous truths.[13] For instance, if it is part of

---

question about the per se accidents of substance (there referred to as τὰ
συμβεβηκότα; 997a26, 29, 33), ignoring the further questions raised in the earlier
passage. The end of Γ 2 (1005a13ff.) treats the per se accidents of being qua
being as the things belonging to substance, but Γ 2 does not actually argue
that what holds good of substances qua beings thereby holds good of substances as such.

12 See my 'Aristotle's Investigation of a Basic Logical Principle,' *Canadian Journal
of Philosophy* **16** (1986), 348 ff.

13 Furthermore, metaphysics will study certain propositions containing both ubi-

the definition of *contradictory* that contradictory terms cannot simultaneously belong to the same subject, then no being can simultaneously possess contradictories. Studying the definition of 'the contradictory' will be equivalent to studying truths about the subjects to which the terms can be applied. Thus to see that the science of being does study the PNC, we should see why it studies the contradictory. Towards this end, let us consider his argument that metaphysics studies various ubiquitous terms.

He begins the argument at 1003b19 ff. with the observation that where there is a single kind (γένος), there is a single (kind of) perception (provided, of course, that one is dealing with a perceptible kind). Since *being* is a kind of subject matter,[14] there is a single science that investigates all of the forms (εἴδη) of being qua being, as well as the various forms of those forms.[15]

---

quitous and (certain) non-ubiquitous terms. For instance, since the primary instance of being is substance, a metaphysical investigation must address the question 'What is substance?' and attempt to discover or establish the general principles that govern substance. One such principle is the *Metaph* Z 6 thesis that each thing that is primary and called what it is called in virtue of itself is identical with its essence. This is a good example of a metaphysical truth that is *not* meant to apply to all beings, and which is neither a definition of, nor a theorem about, an ubiquitous term, though *perhaps* it can be viewed as a theorem following from a number of definitions.

14 Not made explicit here, but see 1004b22 and 1005b9.

15 *Metaph* Γ 2, 1003b19-22. He seems to assume that the kind of science that investigates some kind of thing investigates all of the forms of that thing, as well as all of the forms of the forms. For instance, vowels and consonants are forms of letters, and names and letters are both forms of spoken sounds. Grammar would study all of these.

Alexander, Bonitz and Ross all take the εἴδη τῶν εἰδῶν of b22 to indicate not that the science in question studies the forms of the forms, but rather that various species of the science will study the various species of its subject matter. Applied to metaphysics, this would require that one species of metaphysics studies 'the same,' another 'the similar' and so on. This is not what we find. Aristotle distinguishes various types of sameness, similarity, etc., as part of a continuous exposition. For instance, *Metaph* I 3, 1054a32-b3 distinguishes types of sameness, and b3-13 distinguishes various types of similarity.

His use of the term εἴδη for forms of being is problematic, but nevertheless Aristotle needs some version of the principle that the science that studies the forms of X also studies the forms of those forms in his argument at 1003b34-6 that metaphysics studies 'the same' and 'the similar,' and his argument at 1004a20-1 that it studies contrariety.

He next provides an argument (at 1003b22-34) to the effect that there are as many forms of being as there are of unity.[16] He argues that if 'that which is' and 'that which is one' are the same (1003b22-5), then there are just as many forms of the former as there are of the latter (b33-4). The intervening lines (1003b26-34) provide two bits of support for the antecedent, and hence at b33-4 the consequent is taken as established (ὥσθ' ...). If the conclusion is to follow, the considerations adduced in favor of the antecedent should establish per se (as opposed to per accidens) sameness for 'that which is' and 'that which is one.' The *first* (b26-32) consideration is meant to show that since 'a man,' 'one man,' and 'a man that is' are all the same (b26-7), 'that which is one' is nothing other than 'that which is' (b31-2). Due to a problem with the text at a28 it is difficult to determine how he intends to establish this point. (I think it is likely that he is not drawing an inference, but simply trying to show us that his point is obvious.) The *second* consideration (b32-3) is that the substance of a thing is not one per accidens (and, one might add, is thus a per se unity), as is also (speaking quite generally) the essence of a thing.[17] He does not make it clear how this supports his conclusion. Perhaps the idea is that since unity belongs per se to the being of each being (regardless of whether the being is a substance) it follows that unity is one of the things that belongs per se to the being of each being as such, and hence falls within the scope of metaphysics. In general, both sets of considerations tend to show that unity belongs per se to a being, and is not something external or additional to it. This, together with the principle articulated at 1003b19-22 would suffice to show that the science of being studies the forms of unity. It is much less clear that it actually shows that being has as many forms as unity.

The forms of unity are sameness (or 'the same'), similarity (or 'the similar') and equality (or 'the equal').[18] Consequently, there is a single

---

16 On the interpretation I suggest, b22-a2 is an integral part of a continuous stretch of argument (1003b19-1004a25) for the conclusion that the science of substance studies the terms 'one,' 'many,' 'same,' etc. (see 1004a31-b1). C. Kirwan apparently finds b22 ff. intrusive, writing that it 'interrupts the run of argument' (*Aristotle's* Metaphysics Books Γ.Δ.E [Oxford: Clarendon Press 1971], 82).

17 Here referred to as ὅπερ ὄν τι — precisely some kind of being (for some thing, not necessarily a substance).

18 *Metaph* I 3, 1054a29-31. See also Δ 5,1021a9, where he explicates (i) sameness,

science (namely the science of substance) that investigates the 'what it is' (τὸ τί ἐστι) of each term such as 'the same' and 'the similar' (1003b33-6). Furthermore, since in each case there is a single science for opposites (ἀντικείμενα; 1004a9-10), and since unity and plurality are opposites,[19] metaphysics will also treat of the opposites of the various forms of unity. These are the various forms of plurality: 'the other,' 'the dissimilar' and 'the unequal.'[20] Thus metaphysics studies 'the other' and its forms. One of the forms of 'the other' is 'difference' (διαφορά),[21] and *contrariety* (ἐναντιότης) is a form of difference.[22] Contrariety, being a form of something (difference) that is a form of something (otherness) that is a form of something (plurality) that is the opposite of something (unity) that belongs per se to being, itself must be dealt with by the general science of being.

The outline of the argument to show that metaphysics must study contrariety is this. Metaphysics must study the per se attributes of

---

(ii) similarity and (iii) equality as (i) unity of substance, (ii) unity of quality and (iii) unity of quantity.

Although Aristotle must include both equality and inequality within the scope of metaphysics (since they are forms of unity and plurality), and accordingly lists 'the unequal' as such at 1004a18, nonetheless at 1004b11-12 he lists equality as one of the πάθη peculiar to number qua number — and this despite the fact that he distinguishes the mathematical sciences from metaphysics at 1003a23-6. However, if the equals axiom is one of the common principles (see *Post Anal* A 10, 76a41 and A 11, 77a30-1), then equality should be treated as one of the ἴδια of being.

19 *Metaph* Γ 2, 1004a10 (see also a16-17, excised by Ross; Δ 6, 1017a3 ff; I 3, 1054a20-5).

20 *Metaph* Γ 2, 1004a17-20. This science will also treat of anything called what it is called by reference to one of these, or by reference to plurality and unity. Contrariety is given as an example of just such a further item at a20-2.

21 *Metaph* Γ 2, 1004a21-2. Difference is distinguished from otherness at I 3, 1054b22-31 (see also Δ 9, 1018a12-15). If X and Y differ, then there must be some specifiable respect in which they differ. This need not be the case with otherness. A horse and a man, for instance, are not merely *other* — being in the same genus, they nonetheless *differ* with respect to form. The otherness that differentiates the genus is a διαφορά, or difference (I 8, 1058a5-8).

22 *Metaph* Γ 2, 1004a21 (also I 3, 1054b31-1055a2; I 4, esp. 1055a3-5, 16-17, 22-3; I 8, 1058a8-16). Contrariety is characterized as 'the greatest difference,' or 'complete difference.' Aristotle also points out (answering a question posed at B 1, 995b27) that since contrariety is extreme difference, it is not possible for a contrary to have more than one contrary (cf. 1055b30).

being, one of which is *unity*. There is a single science for opposites, and since plurity is the opposite of unity, metaphysics must also study plurality. However, if metaphysics studies plurality, it must study the various forms of plurality, as well as the forms of those forms, etc. Otherness is a form of plurality, difference a form of otherness, and contrariety a form of difference. Therefore, metaphysics (studying plurality) must study otherness, difference and contrariety.

(c) The science that studies the per se attributes of being treats of contradiction. Although Aristotle does not do so in Γ 2, one can easily extend the train of thought detailed in the previous section to show that the science of being studies *contradiction*.

Contradiction is the primary form of opposition (I 4, 1055b1). A general methodological principle adhered to in metaphysical argument is this: wherever a term has a primary (as distinct from a derivative) application, the science that treats of any of the various applications does so by first studying that to which the term primarily applies, and explains the derivative applications by reference to the primary ones.[23] Contraries are one of four types of opposites, the others being *contradictories* (ἀντιφάσεις), *lacks* (στερήσεις) and *relatives* (τὰ πρός τι, pairs such as master/slave).[24] Since contraries are one type of opposition, the science that studies contraries must do so by first studying contradiction, the primary form of opposition, and then relate contrariety to this, the primary form of opposition.

3 How does metaphysics study indemonstrable logical principles?

The metaphysician must both *state* the (propositional) principles that apply ubiquitously *and* treat of their features. One of those principles will be the PNC. This is a (propositional) principle that applies to each thing that is qua thing that is. One of the tasks of metaphysics is to investigate it, not by attempting to prove that it is true (it is an indemonstrable principle), but rather to investigate truths *about* it. Part II is devoted to this topic.

---

23 Implicit at 1004a25-31 (see also 1003b16ff.).

24 *Metaph* I 4, 1055a38-b1. this is the main topic of *Categories* 10. See also Δ 10, 1018a20-1; I 3, 1054a23ff. (where ἀντίθεσις is used for ἀντικείμενον); I 7, 1057a33-6.

II Proving Things about the PNC

According to *Posterior Analytics* A 2, a first principle is either a common axiom or a thesis. A thesis is either an hypothesis or a definition. An axiom, unlike a thesis, is something such that knowledge of it is required for knowledge of anything whatsoever. A thesis is idiosyncratic to a particular science and as such knowledge of it is required only for knowledge within the confines of that science. For instance, all of the definitions that are utilized within some science are idiosyncratic to just that science, and are not common to any other science. An axiom, on the other hand, is one of the things from which reasoning arises, and knowledge of each axiom is a prerequisite for understanding any proof whatsoever.

At least one way in which the PNC is a prerequisite for proof is that it is a prerequisite for making significant statements. All proof proceeds from premises, and no proof is possible unless the premises are significant and meaningful propositions. One sense in which proof depends upon the PNC is that each proposition in that proof, in order to be significant at all, must conform to that principle. The PNC is common to all reasoning and all proof (ἀπόδειξις) in that it is a prerequisite for the significance of the statements that can serve as premises (or conclusions) of proof. It need not itself occur as a premise in a proof in order for the proof to presuppose it in this manner.

Aristotle claims that the common axioms must be known by anybody who knows anything, and he tells us in Γ 3 (using the verb γνωρίζειν for 'to know') that anybody who is going to γνωρίζειν anything must γνωρίζειν the PNC.[25] The PNC must, in fact, be *better* known than any of the things the knowledge of which presupposes it (see *Post Anal* A 2, 72a25,ff.), and hence nothing could be better known than it. Aristotle commits himself in Γ 3 to even more than this by claiming that the PNC is the *firmest* and most *knowable* of all principles.[26] This goes well beyond what he claims for the other common axioms (e.g., that they must be known by anybody who is to know anything, and that they are better known than scientific theorems).

---

25 Γ 3, 1005b15-16 with b18-20

26 Γ 3, 1005b11-13, 17-23

1 PNC is the firmest of all principles (Metaph Γ 3)

(A) The argument and its assumptions. In *Metaphysics* Γ 3 Aristotle argues that the PNC is the firmest of all principles because it is impossible to be in error with respect to it.[27] To establish the latter, he provides an argument to show that it is impossible to believe a contradiction — that its, it is impossible for a person to at one and the same time believe both members of a contradiction.[28]

The argument for the claim that it is not possible to believe a contradiction makes the following assumptions. First, it assumes the PNC itself as one of the premises. Second, it assumes that beliefs are properties of believers. If Jones believes that Socrates is pale, then Jones has as one of his properties the belief that Socrates is pale. Third is the assumption that the belief in some proposition and the belief in the contradictory of that proposition are *contrary* properties.[29] Fourth, it assumes that if the PNC is true, then it will also be impossible for *contrary* properties to belong to one and the same subject at the same time.[30] Since the Γ 3 proof assumes the PNC as a premise, he can use this fourth assumption to argue that it is indeed impossible for contraries to belong to the same thing at the same time.

Aristotle uses these four assumptions to argue that nobody can believe both members of a contradiction at the same time. It is impossible for contraries to simultaneously belong to the same subject. Beliefs are properties of believers, and the belief that P is the contrary of the belief in the contradictory of P. Consequently, it is impossible for a believer to both believe P and at the same time believe its contradictory.

---

27 Γ 3 1005b8-34. At 1005b13-18 he attempts to support his claim that a principle is the firmest of all if that principle is such that error with respect to it is impossible. He does so by sketching out arguments to show that if error is impossible with respect to a principle, then that principle is (a) the most knowable, and (b) non-hypothetical.

28 Γ 3, 1005b22-34

29 He has in mind here the notion of contraries (ἐναντία) discussed above in Part 1 (2,(B),(ii),(b)), and characterized in *Metaph* Δ 10 as things belonging to the same kind (γένος) that differ as widely as possible within that kind. Thus the belief that Socrates is pale and the belief that Socrates is not pale both belong to the same kind (they are both beliefs), but within that kind they differ from each other as much as is possible.

30 Not argued for in Γ 3, though given a brief treatment later in Γ 6, 1011b15ff.

Were somebody to have *both* beliefs, that person would be in contrary states — but that is impossible (given the truth of the PNC).

(B) A problem with the argument. Rather than discussing all of the various objections that one might make to this argument, here I focus on only one. The objection I have in mind is that the conclusion of the argument is in a significant respect different from what Aristotle should be proving if he is to show that the PNC is the firmest of all principles. If we accept the premises of the argument, then at best we are entitled to conclude that for any believer X, and for any pair of contradictory propositions P and Q, it is necessarily not the case that X both believes that P and also believes that Q. This is what the Γ 3 argument purports to establish. However, Aristotle takes this to show that *it is impossible to be in error with respect to the PNC.* He asserted earlier in the chapter that the firmest principle of all is the one about which error is impossible, and he wishes to show that the PNC is the firmest of all principles by showing that error about it is impossible. That is to say, he seeks to show that one cannot make mistakes about this principle.

The objection I wish to consider makes the assumption that to be in error about the PNC is to believe falsely of it that it is not true.[31] If this is correct, then Aristotle should be arguing that for any believer X, necessarily it is not the case that X believes that there exists a pair of contradictory propositions P and Q such that it is possible that both members of that pair are simultaneously true. However, it is not at all obvious that *this* conclusion is entailed by the actual conclusion of the Γ 3 argument.

To see this, suppose that we accept the claim that no believer can simultaneously believe both members of a contradiction, and consider whether he may nonetheless (compatibly with this supposition) believe that there exists a pair of contradictory propositions P and Q that as a matter of *contingent* fact are not true, but which nonetheless *could* both be true. This believer does not believe both members of a contradiction, but he does believe that it could have been the case that both members of the pair are true. This believer might even claim that

---

31 In (*d*) below I briefly consider a reply that challenges this assumption.

in any counterfactual situation in which both members are true nobody (himself included) believes them both.[32]

The objection continues by observing that if Aristotle wishes to establish that the PNC is the firmest of all principles by arguing that it is impossible to believe falsely of it that it is not true, then the Γ 3 argument fails to provide a conclusion strong enough for his purposes (assuming, of course, that to be in error about the principle is to believe falsely of it that it is not true). Showing that one cannot believe a contradiction does not seem to be the same thing as showing that one cannot believe of the PNC that it is not true, nor does it seem to establish that one must believe that contradictions cannot be true. Even if the Γ 3 argument does show that it is impossible to believe contradictions, it does not show that one cannot disbelieve the PNC, and hence it fails to show that the PNC is the firmest of all principles.[33]

(C) First possible reply to objection. Let us now consider a possible response to this criticism. The reply begins by adopting the thesis that what a person knows, that person also believes, and further assumes that everybody knows something. These two claims are not made in Γ 3. However, we may argue that everybody believes the PNC by combining these two claims with a claim that is made in Γ 3: if one is to know anything at all, he must know that the PNC is true.[34] Hence anybody whatsoever has to know that the PNC is true, and since what one knows, one believes, everybody must also believe it to be true.

Using as a premise the claim that everybody believes the PNC to be true, one can very easily employ the Γ 3 argument (sketched in the previous section) to show that one could not also believe of it that it is not true. The argument is extremely simple: the belief that the PNC

---

32 A further problem (discussed briefly below (in (2)(B)(ii)(b)) in connection with Γ 4) has to do with showing that one must be committed to the principle even for propositions one never actually entertains.

33 It should be noted that the claim that Aristotle is disputing may be formulated as 'one can believe that it is possible for the same thing to be and not be,' and that this is actually ambiguous between (i) somebody can believe that the PNC is not true, and (ii) with respect to some particular pair of contradictory propositions P and Q, somebody can believe both P and Q. The conclusion of the argument is that (ii) is impossible, and the objection is that rejection of (ii) does not entail rejection of (i). Under (*d*) below I discuss the possibility that Aristotle did not intend to argue against (i) in this passage.

34 See note 25.

is true is the contradictory of the belief that the PNC is not true, and since it is impossible to simultaneously believe contradictories, we may infer from the fact that everybody believes that it is true that it is impossible for anybody to believe that it is not.

This defense will have to establish the claim that necessarily everybody believes the PNC is true by considerations that are independent of the argument that nobody can believe a contradiction. If we assume that everybody believes the PNC, we may then proceed to argue that nobody could *also* believe that it is not true.

If this is the strategy that Aristotle is adopting in Γ 3, then (i) he is explaining why the PNC is the firmest of all principles by appealing to the thesis that it is impossible to be in error with respect to it, and (ii) he is explaining why it is impossible to be in error with respect to the PNC by appealing to a set of premises that includes the thesis that anybody who knows anything at all must know the PNC. This latter thesis is not, then, supported by his argument that it is impossible to believe contradictions, and indeed is not supported by anything in Γ 3. In 2 (below) I suggest that it is the purpose of the elenctic argument in Γ.4 to provide support for it.

(D) Second possible reply to objection. A second response to the objection involves *challenging* the idea that to be in error with respect to the principle is to believe falsely of it that it is not true. Perhaps instead, to be in error about a principle (or to make mistakes about it) is to *apply* it incorrectly. If when attempting to give a proof, I misapply some principle, perhaps I may be said to have made an error or mistake with respect to that principle. The mistake need not, and typically does not, consist in disbelieving the principle, but rather consists in a certain kind of lack of competence. In showing that one could never believe a contradiction, one shows that this kind of misapplication is actually impossible when it comes to the PNC.

I doubt, however, that this is what Aristotle has in mind. The opponent of the PNC is somebody who rejects the principle, not necessarily somebody who lacks competence when it comes to applications of it.[35] By claiming that many 'writers on nature' make use of the idea

---

35 Compare: an intuitionist opponent of classical logic is not somebody who lacks competence at applying the law of double negation elimination, but rather is one who rejects the law.

that contradictory statements can be simultaneously true,[36] he is not committing himself to saying that they lack competence at applying the PNC. They cannot (if Aristotle is right) really believe the conclusions of their arguments (although they think that they do), but they are quite aware of the fact that their conclusions violate the PNC.

2 The elenctic proof of the PNC (Metaph Γ 4)

The defense I have suggested (in II,(1)(C)) requires that we have some kind of support for the claim that everybody must know (and hence believe) the PNC. This claim is not something that Γ 3 is concerned to establish, but is, if my suggestion is correct, presupposed by that chapter. The argument I sketched involved taking as premises the statements that everybody knows something, and that if one knows anything at all, one knows that the PNC is true. What reasons might Aristotle have had for finding these premises acceptable? In connection with this question I now turn to a consideration of the elenctic argument of Γ 4.

(A) *Intent of the elenctic proof.* The elenctic argument is supposed to refute an interlocutor who denies the PNC. It does so by getting the opponent to agree to certain premises that actually entail the PNC, or at least particular instances of it. Of course this cannot be proof proper, for proof proper proceeds from premises that are prior to and explanatory of the conclusion, but nothing is prior to and explanatory of the PNC, for it is an indemonstrable first principle. Even were we to have a valid deduction of the PNC from necessarily true premises, the conclusion of which is the PNC, this would not count as a proof proper. Nonetheless, it might be illuminating to see how the PNC follows from premises that even a putative opponent must admit.

Aristotle's intent is to show that adherence to the PNC is a prerequisite for significant thought and discourse. Whether he succeeds is another matter. He is trying to show that it would be impossible to make a significant statement or have a significant thought unless that thought or discourse were in conformity with the PNC. This provides Aristotle with a reason for accepting the claim that everybody must believe of the principle that it is true. Everyone, insofar as he or she

---

36 Consider, for instance, a physicist who mistakenly thought that in order to account for motion he had to allow that Zeno's arrow was, at any given time, both moving *and* at a particular place (and hence at rest and not moving).

is rational, insofar as he or she has significant thoughts at all, presupposes the principle — even somebody so badly confused about his own beliefs as to think that he does *not* believe the PNC.[37]

(B) Problems with the elenctic proof. (i) The elenctic proof makes use of a controversial essentialist principle. I here distinguish what I will call the meta-elenctic argument from the elenctic argument itself. The elenctic argument is an argument directed towards an opponent of the PNC, and it moves from premises accepted by the opponent to PNC as its conclusion. The *meta*-elenctic argument is a reflection on this elenctic argument, and establishes the following claim: if the opponent of the PNC is to signify something, then he will be committed to the PNC governing whatever he has signified in that it would be impossible for the contradictory of what he has signified to be true *if* what he has signified is true. If we can get the interlocutor to admit that he is signifying something, that he is saying something significant, Aristotle thinks that we can then, by getting him to accept further semantical theses about what it is to signify, force him to agree to the PNC. Naturally, an opponent who will not buy into the semantical theses need not accept the conclusion. This I take to be an indication that the project at hand is not to come up with a strategy that is going to refute somebody who is already quite recalcitrant, and even willing to accept contradictions. Reflection on the elenctic argument (i.e., the meta-elenctic argument) is designed to show that the mere possibility of significant thought and discourse requires adherence to the principle. This, in turn, lends support to the claim that everybody has to believe the PNC, for if one did not believe that it governed his utterances (in the sense specified above), he would not be making significant statements.

---

37 There is a question as to whether Aristotle's argument in Γ 4 is intended to establish the claim that everybody *knows* the PNC, or simply the weaker claim that it is believed by everybody. Suppose that he can show that to believe anything, one must believe the PNC. How could he use this to show that if one knows something (anything at all), he must know the PNC? In this paper I do not pursue this question. If Γ 4 is not intended to show that everybody knows the PNC to be true, but is nonetheless intended to show that everybody believes it, that will be sufficient for present purposes. That is, we may now use that conclusion together with the claim that it is impossible to believe contradictories to show that nobody disbelieves the PNC.

This, then, is an independent argument for the claim that everybody must believe the PNC.[38] If one does not, he is no better than a plant.

How does this bring in controversial semantical theses? It does so when it brings in the idea that if a term (say, man) signifies, it signifies *one thing* (σημαίνει ἕν),[39] and the one thing that it signifies will be what being is for man — that is, the essence of man. To say that it signifies one thing, that *man* signifies what being is for a man, is part and parcel with the idea that in the definition of a substance term, what is picked out by the definiendum and what is picked out by the definiens are one and the same.[40] The term 'man' does not signify an item distinct from the essence of man. The content of the definition (the essence) just is what man is, or what it is to be (a) man. He next makes use of the idea that to say that 'man' signifies one thing is to say this: If anything is a man, then *that* (whatever is given in the definition of man) is what being is for a man.[41] This idea is used in support of the claim that if something is a man, then it is necessarily a man.[42]

This last claim is not something that Aristotle thinks follows deductively from his essentialist premises, but rather is something he treats as obvious (just as the syllogism form Barbara is treated as obvious, and not derived from anything else). Take some particular man — Socrates, for instance. Now, if being for Socrates just is whatever is signified by the definition of man, then Socrates could not exist (could not be) without being *that* (whatever that turns out to be). If man is defined as 'biped animal,' then for Socrates being is no other than being a biped animal, and nothing could be Socrates that failed to be a biped animal. We are here going from the essentialist claim that if something is a man, then that is what being is for it, to the modal claim that any particular man must be a man (if he is to be anything at all).

So, how do we get from the essentialist to the modal claim? We are supposed simply to see that it is obvious — it is an obvious case of

---

38 See previous note.

39 Γ 4 1006a28-32

40 See Section IV of my 'On the Origins of Some Aristotelian Theses About Predication,' in J. Bogen and J.E. McGuire, eds., *How Things Are* (Boston: D. Reidel 1985).

41 Γ 4, 1006a32-4

42 Γ 4, 1006b28-30

a predicable belonging to a subject of necessity. We are supposed to see that since being just is (for Socrates) being a man, Socrates would not exist at all (there would not be any Socrates) if he were not a man. This, as I have already indicated, is analogous to what happens in the *Prior Analytics* in connection with the relation 'follows from of necessity' (the relation that relates conclusion with premises in a valid argument). The strategy there was to identify some core, central cases as obvious, rather than attempting to reduce necessity to something else. The kind of necessity with which we are presently concerned, and which is exemplified in such facts as the fact that Socrates is a man, can be thought of as in some sense flowing from the essence of a thing.

Furthermore, just as in the *Prior Analytics* Aristotle distinguishes perfect from imperfect syllogisms (the obvious from the non-obvious), here too we may distinguish cases in which some predicable obviously belongs to its subject of necessity from those in which, although it belongs of necessity, this fact is not obvious. It is obvious that if Socrates is man, then he is necessarily a man (*if* this is what being is for him). An example of a non-obvious necessity would be the fact that Socrates is necessarily receptive of grammar (i.e., capable of learning the grammar of some language). 'Being receptive of grammar' is an ἴδιον, a necessary property, of man in that (necessarily) all and only men are receptive of grammar.[43] 'Being receptive of grammar' is not the definition of man (nor is it part of the definition), but rather is one of the per se accidents of man, and as such is something that should be explicable in the science of man.

We may take this 'theorem' (that man is necessarily receptive of grammar) from the body of science, and combine it with that claim that Socrates is necessarily a man. From these two premises we may validly derive the (non-obvious) necessary truth that Socrates is receptive of grammar. Since Socrates is necessarily a man (since that is what being is for him), and being receptive of grammar is an ἴδιον of man, it follows that Socrates is necessarily receptive of grammar. Although it is not true that being for Socrates is no other than being receptive of grammar, we can show that this property necessarily follows from some predicable (namely, man) that belongs to him of necessity because *it* just is what being is for him.

Part of the task of metaphysics is to isolate certain central necessary truths — not to prove them, but to identify them. Another part will

---

43 See *Topics* A 5, 102a18-30.

be to relate these central necessities to certain other non-central ones in such a way that the latter are explained.

(ii) The elenctic proof is not general enough. (a) A proof is given for substances only. If we were to determine whether the argument given in Γ 4 actually establishes the PNC as a general principle, rather than establishing it for the case of substance predicates only, we are faced with a host of difficulties.[44]

One problem is that the account just sketched applies only to substance terms that give the being of the substance that they pick out. The PNC, however, is perfectly general. It applies not only to statements like 'Socrates is a man,' but also to statements like 'Socrates is pale' (where the item predicated is not any part of the per se being of the subject). In the latter case one cannot argue that Socrates must be pale *because* that (being pale) just is what being is for Socrates, and he would not exist were he not pale. Consequently, we need to find some way to extend the argument so that it applies to propositions other than essential predications.

This is not an objection to Aristotle's strategy, but does point to one difficulty in employing it. Keeping in mind the methodological remarks about metaphysics made in Γ 2,[45] it is not surprising that Aristotle sketches the argument for substance only. Substance serves as the focal point in the analysis of being, and one must first investigate various attributes in connection with their application to substance, and only *after* having done this does one extend the investigation to the various items that are ontologically dependent upon substance. In Γ 4 we get the elenctic proof for the case of substance predications. Unfortunately, Aristotle never gives us an extension of the proof to the non-substance cases.[46]

(b) A proof is given for believed propositions only. A second, and to my mind more serious, difficulty is that at best the elenctic argument establishes a kind of limited belief, or adherence to the PNC — namely,

---

44 I wish to stress that the sense in which Γ 4 is meant to establish the PNC is not to give a proof proper, but rather to give an elenctic proof which can serve as the object of reflection by the meta-elenctic proof.

45 See I, 2, (B), (ii), (c) above.

46 One might start by observing that whenever a non-substance predication of the form 'S is P' is true (where S is a particular substance), there will be an accidental being, 'the P,' such that (i) 'the P' is accidentally the same as S, and (ii) being for 'the P' is no other than being a P.

any time one attempts to make a significant statement, or have a significant thought, he is thereby committing himself to acceptance of the PNC in connection with that particular statement (or thought). This is not the same as showing that anybody engaged in significant statement-making must believe the PNC in full generality. A person might believe that the PNC does govern his significant thoughts and statements, and yet *disbelieve* the general principle. He might even believe that there are all sorts of true contradictions, but that a limitation of the human intellect prevents us from knowing them. Perhaps such a person is deeply confused, but nonetheless the fact that he must regard each of his significant statements as governed by the PNC (if it is a fact) does not by itself show that he must regard the PNC as true in its full generality. So even if Aristotle's position is correct, much more needs to be said than what he provides in Γ 4.[47]

---

47 I am grateful to Mohan Matthen and the other participants of the *Aristotle To-day* conference for assistance in clarifying the issues. Also I received helpful advice when I presented a previous draft to my seminar at UCLA and at the 1986 Berkeley Colloquium in Ancient Philosophy, and would in particular like to thank David Blank and Michael Griffith. Finally, Glenn Most made valuable comments that aided in the transition to the final draft.

Individual Substances as Hylomorphic Complexes[1]

I Introduction: Aristotelian Forms as Causes

Aristotle held that for each natural thing there is something that constitutes being that thing — call this its *essence* or *form*.[2] And he says that nothing could fail to possess the essence characteristic of its kind (see for example *Metaph* 1006b32-3). It follows that for a thing to cease to possess its essence is for it simultaneously to cease to exist — for instance, for a human to lose the soul characteristic of humankind is for that human to die. This implication offers us a justification for saying that, according to Aristotle, essential properties are necessary properties.

Recently, Aristotelian scholars have made much of this connection between essences and necessity, because they have hoped to exploit the tools of modern modal logic in order to make sense of Aristotle's metaphysical ideas.[3] Thus it has been suggested that as a first approximation Aristotelian essences be *defined* as consisting of necessary properties.

---

1 I am greatly in debt to Martin Tweedale for extensive discussion of an earlier working paper, and to my research assistant Steven DeHaven for constant criticism and help. I also acknowledge a grant from the Social Sciences and Humanities Research Council of Canada that helped make this work possible.

2 Aristotle calls it variously the 'what-is-it' (τὸ τί ἐστιν), the 'what-it-was-to-be' (τό τι ἦν εἶναι), the 'form' (εἶδος), the formula (λόγος), and the substance (οὐσία). Any differences among the uses of these terms is irrelevant for my present purposes.

3 See for example Alan Code, 'Aristotle's Response to Quine's Critique of Modal Logic,' *Journal of Philosophical Logic* **5** (1976) 159-86. Code's views have changed since he wrote this article, and I do not know whether he still thinks that recent modal semantics shed light on Aristotle's essentialism.

But this exegesis cannot be complete, for it fails to take account of an important complication. Aristotle says that a thing possesses its essence 'in virtue of itself' (καθ' αὐτό), or intrinsically. But not all the properties a thing possesses intrinsically are part of its essence: Aristotle also allows for what he calls 'intrinsic accidents' (καθ' αὐτὸ συμβεβηκότα), or 'propria' (ἴδια),[4] which are *not* part of a thing's essence, but belong to a thing because of that thing's essence. (An example is the property of having angles that sum up to two right angles: this is not part of the essence of a triangle, but belongs to triangles because of their essence.) Thus Aristotle conceives of essences as *core* properties, and he holds that they determine some of a thing's other properties.

It is not enough, then, to understand essences *simply* as consisting of necessary properties; we should also try to account for their priority with respect to other properties. If one supposes, as I do, that Aristotelian metaphysics is to be understood *ontologically*, there seems to be only one route to an adequate interpretation of the priority of essences, and that is to make them *causally* prior.

Given this conception, and following up on the idea of interpreting Aristotelian essentialism in conformity with modal logic, one might now think of defining intrinsic accidents as the properties a thing has as a logical consequence of possessing core essential properties. And this treatment is indeed strongly indicated by the framework Aristotle erects for scientific exposition in the *Posterior Analytics* — there, all sorts of causal links are incorporated into the axiomatic exposition of a science.

But I do not think that in the end this sort of approach does justice to Aristotle's conception of substances that admit change. For it would seem that Aristotle wants to allow that there can be interfering factors that foul up the relationship between essence and consequent attributes. As a result of these interferences, the 'intrinsic accidents' may not be strictly universal across a species, but may instead appear to different degrees, or in different ways, or even fail altogether to be instantiated. This failure of universality blocks the explication in terms of necessity, for obviously if it is not the case that all Fs are G then it cannot be necessary of all Fs that they be G — indeed, given that

---

4 In the *Topics*, 102a18-20, the *idia* of kind K not only belong necessarily to all Ks, but are 'convertible' as well, that is belong only to Ks. I am not concerned here with this additional constraint.

the essence common to *F*s is what accounts for their necessary properties, it cannot even be true of *any F* that it is necessarily *G*.

The idea that things of a kind may differ from one another in this way goes back to Plato. For Plato, the forms do not (like *mere* universals) simply reify the similarities that (already) exist among the members of classes of particular — Justice, for example, is *not* thought by Plato to be simply that which is shared by all just things. Rather, forms are ideals toward which particulars strive, *always* unsuccessfully. The forms can exist only in an atemporal domain, says Plato in the *Timaeus* (37-40), and in making copies of these forms suitable to reside in Time, the Demiurge is forced to copy the forms imperfectly. Hence the definition that constitutes a form will not be found perfectly instantiated in the things that participate in the form. In dialogues like the *Phaedo*, the *Republic*, and the *Symposium*, Plato spends time elaborating on the ways in which particulars may fall short of the forms: they are relational while the forms are absolute, they are temporary while the forms are atemporal, and so on.

Aristotle's forms (or essences) are not quite like Plato's: they are not ideals, unattainable or otherwise. Nevertheless, the properties that they explain also admit of being instantiated in degrees or in different ways. For example, slaves and women are humans, according to Aristotle, and as such they possess the essence or form of humanity (*Metaph* X 9). But they do not, to the same degree or in the same way, possess the rational faculties that (according to *de An* II 2) are associated with the human form (*Pol* 1260a12-17). A biological example of the same phenomenon is this: the reproductive capacity is definitive of humans and part of their form (as it is with all living things). Still, the reproductive capacity is found in different individuals in different forms (males and females are the most obvious example), and to different degrees. What is more, these variations are *explainable*: it is not the case that they are treated *merely* probabilistically, as some treatments of properties that hold 'for the most part' would suggest.[5] In these cases, Aristotle is interested not only in properties that obtain invariably across a kind, but also in properties that obtain *normally*, i.e., in the absence of the sort of factors that would *explain* departures from the norm. Thus, when Aristotle says that the explanatory framework of the *Posterior Analytics* is applicable not only to properties that hold invariably of members

---

5 For the notion of a property that holds 'for the most part,' see *Pr Anal* 32b5-10, *Metaph* 1064b30-65a5, *Post Anal* 43b32-8, and *Phys* 198b5-7.

of a natural kind, but also to properties that hold 'for the most part,' we should not take him simply as 'introducing a quasi-modal operator' — as introducing a frequency operator weaker than universality.[6]

Now it seems that the notion of an essence as a necessary property is not, by itself at any rate, going to help us to understand either why properties associated with essences are instantiated to different degrees or why they do not hold universally — the notion of a necessary property is not rich enough to have any relevance to degrees, and it is simply incompatible with the failure of universality. Aristotle himself seems to treat these cases as failures of essences (i.e., forms) as causes. Consider his views about embryonic development. Here we find that various abnormalities ranging from birth-defects, to the failure of a male child to resemble its father, to occurrences that even Aristotle must have considered normal — the birth of a female child, for example — are explained by a failure of form to master recalcitrant matter, or by the operation of interfering causes.[7]

When we think of such cases, we begin to see why we should think of Aristotelian forms, and of Plato's forms, as *causes*. Suppose, then, that possessing a certain sort of essence is constitutive of being a member of a particular kind, and that therefore having that essence is necessary for things of that kind. But then let the relationship between the possession of the essence and the possession of properties outside the core be causal — that is, let the essence tend to cause things to have these other properties, provided that the conditions are right.[8] A thing might then fail to have some of the properties associated with its form.[9]

---

6 I am quoting here from Jonathan Barnes' commentary on the *Posterior Analytics* (Oxford: Clarendon Press 1975), 184. Barnes concedes that 'Aristotle himself never worked out a satisfactory logic for "for the most part propositions."' I would go further: he did not have the slightest conception of what such a logic would even look like. Surely this substantiates the point that a merely frequentist analysis is not what he had in mind.

7 I am indebted to Joan Kung for suggesting to me the relevance of this sort of example.

8 This idea can be found, for example, in Sarah Waterlow's book *Nature, Chance and Cause* (Oxford: Clarendon Press 1982). See chapter 1 in particular.

9 St. Thomas Aquinas gives a succinct and accurate account of how causes can interfere with forms: 'As regards the *individual nature*, woman is defective and misbegotten, for the active power in the male seed tends to the production of a *perfect likeness* according to the masculine sex; while the production of woman comes from a defect in the active power, or from some material indisposi-

When he articulated the modal aspect of his theory, Aristotle no doubt had in mind the normal case where nothing prevents the essence from being expressed. In the normal case, the presentation of Aristotelian essences in terms of necessary properties may not be misleading. Even so, it is important to remember that this presentation does not get to the heart of the Aristotelian notion, and that even this 'normal case' should be regarded as a reflection of the deeper conception of forms as causal factors.

My aim in this article is to present an ontological framework for Aristotle's theory of essence which is capable of accommodating its causal role. The interpretation I shall offer is supposed also to accommodate an Aristotelian dictum that makes it difficult to make an essence any sort of *property*, necessary or otherwise: namely that an individual is identical with its essence.[10] (Since Aristotelian individuals are not properties, it follows that essences are not properties either.[11]) As such, my interpretation attempts to solve a textual problem that has always faced interpreters of the central books of the *Metaphysics*. *Metaphysics* Z begins with the assertion that *primary being* is 'τί ἐστιν καὶ τόδε τι' — the 'what-it-is *and* this' (1028a11-12). The treatise goes on to argue the claims of *form* or *essence* (which is presumably to be identified as the what-it-is) to be considered as primary being (1032b1, 1033b17, 1037a5, 1037a29, 1041b7) without ever giving up, and sometimes seemingly endorsing (1035b28, 1040b23, 1041a3) the thesis that primary being is a this, i.e., an individual.[12] There is, apparently, a

---

tion, or even from some external influence such as that of a south wind, which is moist as the Philosopher observes' (*Summa Theologica* I.92.1).

10 A proper defence of this claim would need to discuss the relationship between the form of an individual, the concrete whole composed of form and matter, and what I call the 'manifold' of all the properties a thing possesses (p. 163 below), with special reference to the following passages in *Metaphysics* Z and H: 1031b28-32a5, 1036a16-25, 1037a5-9, 1037b4-7, 1043b2-4. An interpretation opposed to mine will be found in Alan Code, 'Aristotelian Theses About Predication,' in J. Bogen and J.E. McGuire, eds., *How Things Are* (Boston: D. Reidel 1985).

11 Of course, some people have drawn a different conclusion here: suggesting that the primary substances in the central books of the *Metaphysics* are not individuals, thus implying that there is a major divergence between this work and the *Categories*. It is a virtue of my approach that it becomes unnecessary to posit any such divergence.

12 *Metaphysics* Z 13 is particularly important here, for it is argued there that 'what

contradiction here. What-a-thing-is is universal: for example, what-you-are and what-I-am are the same — human (or so one should gather from *Cat* 2b7-12). How is it possible that primary substance should be both a *this* or individual, and *what-it-is* or universal?[13] Mine is not the only tenable answer to this question, but it is a solution: there are individual instantiations of the what-it-is, and these are causes.

My interpretation is of some contemporary interest, I think, for three reasons.

1) Some recent authors — for example, Armstrong, Tooley, and Dretske[14] — have revived the view, which I think was held by both Plato and Aristotle, that explanatory generalizations, or laws, have *universals* as their domain. This view is, on the face of it, in conflict with another view commonly held today, for example by Wesley Salmon and Nancy Cartwright,[15] that only causal relations can be be genuinely explanatory; for most people believe that universals are not causally efficacious. Now this is, I think, just the sort of problem that Aristotle faced when he tried to relate the explanatory scheme of the *Posterior Analytics*, which deals with 'laws' conceived of as having universals as their domain, to the scientific treatises, which deal with causes that relate concrete individuals.

Now my interpretation of Aristotle's theory does offer an answer to this problem. (How satisfactory the answer is I shall not discuss.) According to my interpretation, Aristotle embraces a version of the view

---

is said universally cannot be substance; for primary being is that which is peculiar (ἴδιος) to the individual' (1038b8-10), and that 'no universal attribute is a substance, and no common predicate indicates a "this," but rather a "such."'

13 See James Lesher, 'Aristotle on Form, Substance, and Universals: A Dilemma,' *Phronesis* **16** (1971) 169-78.

14 See Michael Tooley, 'The Nature of Laws,' *Canadian Journal of Philosophy* **7** (1977) 667-98; Fred Dretske, 'Laws of Nature,' *Philosophy of Science* **44** (1977) 248-68; D.M. Armstrong, *What Is a Law of Nature?* (Cambridge: Cambridge University Press 1983); and David I. Lewis 'New Work for a Theory of Universals,' *Australasian Journal of Philosophy* **61** (1983) 343-77.

15 Wesley Salmon, 'Why ask "Why?"?,' *Proceedings and Addresses of the American Philosophical Association* **51** (1978) 683-705, and *Scientific Explanation and the Causal Structure of the World* (Princeton, NJ: Princeton University Press 1984); and Nancy Cartwright, 'Causal Laws and Effective Strategies,' *Noûs* **13** (1979), reprinted in her *How the Laws of Physics Lie* (Oxford: Clarendon Press 1983), essay 1.

that individual substances are 'bundles' of individual attributes. The universals that are the subject of laws are instantiated in these attributes, and causal relations are on the domain of the individual attributes.

The view that substances are bundles of attributes is held also by Hume, for example. It is a valid criticism of most versions of this view that it makes individuals into complex attributes, and thus negates the category distinction that ought to be maintained between individuals and attributes. Now of course Aristotle does maintain that there is a category distinction between individual substances and attributes. And I shall argue that there is an idea implicit in his philosophy that makes this category-distinction consistent with the idea that individuals are bundles of attributes.

2) Aristotle's position, as I interpret it, is *actualist* with respect to essences. That is, the difference between an essential property of a thing and one of its non-essential properties is one that manifests itself in the actual world, and not merely as a result of differences that hold in non-actual possible worlds. This is desirable because Aristotle's essences, on my interpretation of them, are causal entities. Thus it cannot be that the constitutive differences between essential and non-essential properties should be non-actual, since non-actual differences cannot account for differences in causal roles.

Actualism with respect to the explication of essences is something that could, I think, be of some value to contemporary essentialists as well. The present custom of explicating essences by reference to conditions in non-actual possible worlds notoriously gives modal metaphysicians a choice between two embarrassing positions: either they must hold that the unreal can make a difference, or that the non-actual can be real. The actualist position on essences offers us a third choice, that of not referring to possible worlds at all.

3) Contemporary essentialism has to face the difficulty that many properties considered characteristic of a sort of thing do not in fact universally belong to things of that sort. For example, not all water is tasteless, not all humans are rational, not all gold is yellow, and so on. As I have said, I take Aristotle's essentialism to be directed towards explaining normal as well as necessary properties.

II Individual Attributes

I begin by recalling the Aristotelian idea that there are individual attributes; for it is against the background of such entities that I think that the question 'What is substance?' is best understood.

An individual attribute is, first of all, a member of a category other than substance — it might be a quality such as a color or a virtue, or a quantity such as a surface or a line.[15a] It is secondly not a universal. In other words, it is possible to have exactly similar but numerically distinct instances of color, shape etc. — and these are the individual attributes.[16] G.E.L. Owen has argued that there are no individual attributes recognized in Aristotle's philosophy.[17] (The position is a very influential one, and was warmly endorsed by Montgomery Furth at the conference.) Owen's argument is curious, because it is, in the main, a series of attempts to show that this Aristotelian passage or that is capable of being interpreted as not being committed to such entities — in particular *Cat* 2. But most of these passages are in fact compatible with individual attributes, admittedly sometimes with an implausible degree of elaboration.[18] In any case the texts have been thoroughly worked over without yielding a resolution of the controversy. The case Owen makes is really driven by a scepticism that individual attributes do any work, or that they even make sense: so I think that the best way to counter Owen is to put the *Categories* texts aside and show instead why Aristotle's system cannot do without individual attributes.[19]

15a It is not only determinate universals that will have individual instances. An indeterminate universal like *color* will be 'said-of' some more determinate universal, like *scarlet*, until we arrive at a completely determinate universal. Any instance of such a completely determinate universal is simultaneously an instance of the less determinate universals said-of it — in other words, an individual scarlet is simultaneously an individual red, and an individual color.

16 Recent discussions of entities of this type originate with G.F. Stout, 'The Nature of Universals and Propositions,' *Proceedings of the British Academy* (1921). More recent versions include D.C. Williams, 'On the Elements of Being,' *Review of Metaphysics* 7 (1953) 3-18; Wilfrid Sellars, *Science, Perception and Reality* (London: Routledge and Kegan Paul 1963); and K.K. Campbell, 'The Metaphysics of Abstract Particulars,' *Midwest Studies in Philosophy* 6 (1981) 477-88.

17 G.E.L. Owen, 'Inherence,' *Phronesis* 10 (1965) 97-105

18 See James Duerlinger, 'Predication and Inherence in Aristotle's *Categories*,' 15 (1970) 179-203.

19 For an excellent discussion of Aristotelian texts other than those in the *Categories*, see Robert Heinamann, 'Non-Substantial Individuals in the *Categories*,' *Phronesis* 26 (1981) 295-307.

Appropriately enough, the case I shall make has to do with Aristotle's theory of causation. Given my subsequent treatment of essences, the case I make has the direct consequence that essences must have individual instances. These are, according to me, the Aristotelian individual substances.

The notion of an individual attribute is, I think, a by-product of the spatial nuances of some Greek ways of dealing with predication. Many Greek philosophers – Anaxagoras, Protagoras, Plato, and Aristotle are examples that come to mind – analyzed 'Socrates is white' in terms of whiteness being spatially or quasi-spatially related to, being present *in*, Socrates. (See for example *Phys* 210a27-b1.) But now suppose that Socrates and Coriscus are characterized by *exactly* the same shade of white. Then if there were no individual attributes to mediate the relationships between Socrates and Coriscus and the determinate universal they share, it would follow that the universal is in two different places at the same time. But this leads to the dual absurdity of supposing (1) that a universal is in a place, and (2) that numerically one and the same locatable thing can be in two places at once.

The only way to avoid this difficulty while retaining the spatial overtones of the theory of inherence is to distinguish the two individual instances of white that inhere in Coriscus and Socrates. These may then be related to the universal in some appropriate way. For example, suppose one held that there is an individual whiteness in Socrates, and a numerically distinct but qualitatively identical whiteness in Coriscus. Universal whiteness might then be held to be a composite of all such individual whitenesses – as in the sail analogy of the *Parmenides*.[20] Or better, the universal could be conceived of as an abstract entity outside space and time, with spatio-temporal individual instances *in* the different things characterized by it. This was the position Plato took in the *Phaedo*, and I believe it was Aristotle's position.

Individual attributes have an indispensable role to play in Aristotle's theory of causation as well. The essential idea here is that Aristotle's theory demands that non-substantial items be causes. And, as does

---

20 Another example is pseudo-Joscelin in *De Generibus et Speciebus*, Peter King, ed. (1982 Princeton University dissertation, 'Peter Abailard and the Problem of Universals'), 128-42: 'I say that humanity inheres in Socrates not because the whole is consumed in Socrates but because only one part of it is informed by Socrateity. For I am said to touch a wall not because each part of me sticks to the wall, but perhaps only the tip of my finger.' (I am grateful to Martin Tweedale for this reference.)

any adequate theory, his allows that individual items may be caused. But individuals require individual causes. He must therefore allow that there are individuals in non-substantial categories.

So consider Aristotle's explanatory framework for κίνησις, or change, in the first three chapters of *Physics* III. We are told here that there is an active potentiality in the mover, a passive potentiality in the moved. When the mover makes contact with the moved, the potentialities are actualized, and there is movement in the moved – this movement being simultaneously the actualization of both potentialities. We have here three things that are said to be *in* a subject – the two potentialities and their mutual actuality. These three things have to be individuals. For suppose that qualitatively the same potentiality to teach is in me and in Pelletier, and that qualitatively the same potential to learn is in Mavko and Wylie. Qua universals these potentials are the same, so if it were *universals* that came into contact and became actual, I could teach Wylie simply by having Pelletier teach Mavko. An economical procedure, no doubt, but unfortunately it is an impossible one. Similarly, one might argue, the 'perceptible forms' in Aristotle's account of sensation, in *de An* II 12, must be individual. Individuals require individual causes.

Aristotle clearly recognizes these exigencies of his theory. In *Physics* II 3 causal talk can be generic or particular (195b13-14), and he says:

> Actually operative (τὰ ἐνεργοῦντα) particular causes exist and cease to exist simultaneously with their effects; for example, the healing (ὁ ἰατρεύων) with the being healed... But this is not always true of potential causes; the house and the builder do not pass away simultaneously. (195b17-21)

Obviously, he is talking about actualized potentials here – it is in no sense the case that the doctor perishes immediately after he has finished administering a treatment, only to be resurrected when he begins to deal with his next patient. It is an instance of actualized doctoring that perishes after the treatment is completed. This thing, the actualized doctoring, has to be an individual – universals never perish. Aristotle indeed says as much: he is talking about particular (καθ' ἕκαστα) causes.

The only way to evade the force of the argument in the last paragraph is to deny that items in categories other than substance are terms in causal relations. Such a denial could be extrapolated from a claim recently made by Alan Code:

> Soul is that in virtue of which ... we live, think, perceive, and so on.
> *We* think, for instance, *in virtue of* the fact that our bodily parts are formed,
> structured and organized with that *kind* of organization that a set of bodily
> parts must have in order that the constituted individual can think. (10,
> my emphasis)[21]

Perhaps, following Code, someone would want to analyze the truth-conditions of this sentence:

> Pelletier's potential-to-teach acts upon Mavko's potential-to-learn.

in terms of this one:

> Pelletier acts upon Mavko in virtue of his participation in a cer-
> tain universal active potential and Mavko's participation in a cer-
> tain universal passive potential.

Since the paraphrase envisages only individual substances as terms in causal relations, the force of my argument in favour of introducing individuals in other categories is nullified.

I have not the space here to take issue with this strategy in any detailed kind of way. I shall only say that even if the paraphrase is one that Aristotle would have accepted as correct, his writings on cause and change lead me to believe that he would have cashed out the 'in virtue of' clause in the paraphrase in terms of individual attributes. So, for example, he says in *Phys* II, 195a32-5, that 'the being of Polycleitus' is a cause of a statue by supervenience, because it supervenes upon 'the sculptor.' And at *de An* 418a21-3, he makes a similar point, saying that 'That which you perceive [e.g., Diares' son] supervenes on the white thing,' the latter being the proper object of vision. These statements suggest the priority of the first locution above, for the purposes of explaining causation, for they suggest that instead of saying something like 'Polycleitus made the statue in virtue of being a sculptor,' Aristotle would rather say that the art of sculpting is directly responsible for the statue, and Polycleitus only indirectly, by virtue of supervenience.

---

21 Alan Code, 'The Aporematic Approach to Primary Being in *Metaphysics* Z,'
    Canadian Journal of Philosophy Supp. Vol. X (1985) 1-20

It is important to note that the notion of individual attributes out-
lined above is perfectly compatible with there being two individual
attributes that are qualitatively identical but numerically (i.e., spatio-
temporally) distinct. (Indeed the notion of an individual attribute is
motivated by this possibility.) This will become important later; for I
shall attempt to make the relationship between form and matter an
instance of that between attribute and substance. (Form, I shall say,
is predicated of matter in just the way in which a non-substantial
attribute is predicated of a substance.) It follows that there will be individual forms, but it does not follow that these will be qualitatively
different. Thus it will not be the case that for every pair of distinct individuals, there will be a pair of qualitatively distinct forms — there
will however be a pair of numerically distinct forms for every pair of
individual substances.

III How Should We Understand the Question, 'What Is Substance?'?

It is now possible to give meaning to the question: 'What is substance?',
which Aristotle asks at the beginning of *Metaphysics* Z. Envisage the
following set of individual attributes — the colors, sizes, virtues, capacities, etc. that are mine. (This is not a good characterization of the
set, because it is question-begging: it assumes that attributes belong
to a substance. But in the absence of the capacity to point to the set,
in a written paper, it will have to do.) Intuitively, the set has a unity:
there is something that binds its members together. One indication
of this, noticed by Plato in the *Phaedo*, is that it is not possible to combine contrary attributes into a set of this sort:

> It seems to me that not only is largeness itself never willing to be large
> and small at the same time, but also that the largeness *in us* never ad-
> mits the small, nor is it willing to be overtopped. Rather, one of two
> things must happen: either it must retreat and get out of the way, when
> its opposite, the small, advances towards it; or else, upon that oppo-
> site's advance, it must perish. But what it is not willing to do is to admit
> and abide smallness, and thus be other than what it was. (102d, Gal-
> lop's translation, my emphasis)

Plato tries to convey the idea just in spatial terms, but he realizes that
this is not going to work. The exclusion of the small by the large is
simply not governed by closeness: my two-year old just doesn't get
any bigger when I give her a hug. That is why Plato has to sneak in

the largeness *in us* specification: the point then is that a quality *in something* excludes its contrary *in the same thing*. (Cf. Aristotle, *Metaph* 1007a21: those who deny the principle of non-contradiction 'do away with substance and essence.' The point is not the same one, but is related.)

We cannot, then, do just with sets of attributes; these sets have to be unified into *manifolds*, which are predicated of a single thing. The question is: how shall we identify a set of attributes as a manifold, or in other words, as being a set of attributes that is predicated of a single thing? *Perceptually*, all that unifies a manifold is its spatial unity, and as we shall see, Plato tried to parlay this spatial unity into a theory of the subject of predication. He failed. Aristotle's attempt is, I shall argue, to identify a sub-set of the attributes in a manifold with the subject of predication – with a substance. The question 'What is substance?' is, according to my interpretation, a question about how to identify this subset.

Now, we know that in the *Categories*, Aristotle divides the attributes of a thing into those that are 'said-of' it, and those that inhere. This distinction enables us to refine the problem a bit further. Look at it like this: we are presented with manifolds of individual attributes, and in each such manifold we know that some of the attributes constitute the entity in which the others inhere: in other words, the manifold is divided into the attributes that inhere, and those that constitute the *receptacle* in which they inhere. But there is no phenomenological difference corresponding to this distinction between the attributes that inhere and those that constitute their receptacle.[22]

The question 'What is substance?' concerns the constituents of the receptacle, or substance. Which of the individual attributes present in the manifold are part of the unity that constitutes the substance, and which are not? The aim of Books Z and H of the *Metaphysics* is to state generalities concerning questions of this form. As Aristotle states the problem in Z 1, the problem is to distinguish between ways of identifying a substance. One way is by identifying the non-substantial attributes that are associated with it – for example, by saying 'the substance there that is good, or three cubits long.' But this

---

22 In fact, any phenomenological distinction is apt *not* to give us intrinsic properties, since what is 'prior to us' (what seems to perception and the untutored mind to be most important) is different from what is 'prior in fact' – see *Phys* I 1; *Post Anal* 71b34-72a6; etc.

way is relational, it is like identifying a person by identifying his or her offspring, and it does not count as telling us what the substance is 'in itself,' nor as helping us to 'know it most fully.' To know a substance most fully, one must know which of the attributes in its manifold constitute it — one must identify it by its own parts, not merely by the things that inhere in it.

My account of the question 'What is substance?' would not be available if substances were not constituted by individual attributes. For if there were none, what would the contrast be between 'the thing over there that is white' and 'the thing over there that is a man'? Both identify a substance by reference to a property it possesses. The distinction between the two ways of characterizing a substance is now unmotivated.

So the smoothest way to read Aristotle is to think of a *collection* (or manifold) of individual things — the man, his 'how' or qualities, his 'how-much' or shape, size, etc., his 'where,' and so on. Each of these things may be characterized *extrinsically*, by means of other individuals in the collection — 'the color Coriscus has,' 'the place in which the substance that has paleness is' are examples of such characterizations. But we know each of these things best when we are able to characterize them *intrinsically* — by means of its components, and this comes to 'by means of its definition' (Z 4).[23]

This exposition gives a good, and literal, account to the sense of the term κατὰ συμβεβηκός (= *per aliud*, = extrinsic) that is set forth at *Metaph* Δ, 1025b30ff. For that is a sense of 'by accompaniment' that is supposed to be compatible with necessary, or invariable, belonging — yet it excludes things that are 'part of the substance.' Thus having angles that equal two right angles is essential to triangles κατὰ συμβεβηκός — this is because what is constitutive of triangles is not this property but being bounded by three sides. My proposal has the consequence that we should take the term κατὰ συμβεβηκός, and its correlative καθ' αὑτό (= *per se*, = intrinsic), as being used in this way in the central books of the *Metaphysics*, not as meaning 'accidental' and 'essential.' (See note 8.)

These definitions carry the important benefit that we can reconcile the terminology used in the *Categories* with that used in the *Metaphys-*

---

23 As it turns out there are complications in characterizing non-substantial things intrinsically, and these lead Aristotle to say that the 'what' applies only to substances in the primary sense (*Metaph* 1030a29-32). But these complications are not relevant to the point being made here.

ics. The argument is a little complicated, however, and readers to whom this benefit does not mean much may skip to the next section.

There will be, as a result of the foregoing remarks, two kinds of predication to deal with in *Metaphysics* Z and H, and these will not correspond exactly to necessary and accidental predication.

1. *P* is predicated καθ' αὑτό of *S* if an individual instance of *P* is a constituent of *S* (or if *S* is an universal: an individual instance of *P* is a constituent of any individual instance of *S*).

For example, *color* is predicated καθ' αὑτό of *white* because a *color* is part of what constitutes any individual instance of *white*, and *animal* is predicated καθ' αὑτό of Jones because *animal* is part of what constitutes Jones. Here, we shall understand that when *F* is part of what constitutes *G*, an individual instance of *F* is a part of every *G*. (This involves some regimentation: it is clear that in some places, especially in the *Categories*, Aristotle speaks as if an individual instance of *color* is identical with, and not merely a part of an individual instance of *white*, and similarly as if Jones were identical with an individual instance of *animal*. But I would claim that when he comes to elaborate on the metaphysical, as opposed to the merely logical, analysis of such things he takes the view that I am here outlining.)

This leads to a second definition:

2. *P* is predicated κατὰ συμβεβηκός of *S* if an individual instance of *P* is associated with *S* in the same 'manifold' without being a constituent of *S*.

Color is predicated κατὰ συμβεβηκός, not καθ' αὑτό, of man because colors only accompany men — men are not constituted by color.

Now add to these definitions the following:

3. *X* is an individual instance of *Y* if and only if *X* is individual and either *X* or a constituent of *X* satisfies the definition of *Y*.

and we have the consequence that where *P* is predicated of *S*, *S* will satisfy the definition of *P* if and only if *P* is predicated of *S* καθ' αὑτό. If *P* is predicated of *S* κατὰ συμβεβηκός, then the individual instance of *P* merely accompanies *S*: *S* is not the instance itself. If, on the other hand, *P* is predicated of *S* καθ' αὑτό, then the individual instance of *P* is a part of *S*, and this means that *P* is part of the definition of *S*.

This is an important result, because it brings the relations of predication καθ' αὐτό and κατὰ συμβεβηκός, in the special sense that we are dealing with here, into line with the distinction between the said-of and inherence relations made at *Cat* 2a19-34. (It is a particular virtue of the above account that it explains the peculiar status given to the differentiae of substances in the *Categories*, for it is said there that the differentiae are not in the category of substance, but are nevertheless said-of substance [3a21-9]. This is inexplicable on most accounts, but follows directly from mine, since differentiae are parts of the definition.)

IV The Platonic Origins of the Theory of Substance[24]

Having thus attempted to clarify the question 'What is substance?', I turn now to Plato's attempt to answer it. (I do not, of course, claim that Plato used Aristotelian terminology.)

The first really clear understanding of substance and its role vis-à-vis individual attributes occurs, I believe, in Plato. To illustrate this, let us consider the theory of predication that forms the framework for Plato's exposition of Protagorean relativism in the *Theaetetus*. The crucial passage reads as follows:

> When the eye and some appropriate object which approaches beget whiteness and the corresponding sensation — which could never have been produced by either of them going to anything else — then when sight from the eye and whiteness from that which helps to produce the color are moving from one to another, the eye becomes full of sight and begins at that moment to see, and becomes, certainly not sight but a seeing eye, and the object that joined in begetting the color is filled with whiteness and becomes in its turn not whiteness but white. (156d-e, cf. 159d-e: Fowler's translation)

The main points of Plato's presentation are as follows:

First, a certain correlation is made between abstract nouns like 'whiteness' and 'sight' and the corresponding adjectives 'white' and 'seeing.'

---

24 The view of Plato being developed in this secton owes much to John Driscoll, 'The Platonic Ancestry of Primary Substance,' *Phronesis* **24** (1979) 253-69. The discussion of the *Theaetetus* is, however, my own: see my 'Perception, Relativism, and Truth: Reflections on *Theaetetus* 152-160,' *Dialogue* **24** (1985) 33-58.

To wit, the claim is that certain independently existing entities, which we might call *subjects* come to be characterized by an adjective when the items designated by the corresponding abstract noun come-to-be, and enter their subjects. Thus we have:

> When a whiteness comes to be, 'the object that joined in begetting it, be it a stick or a stone' becomes white.

> When a sight comes to be, the eye becomes seeing.

Second, the above correlation is joined to a theory of inherence: the sight and the whiteness 'fill' their respective subjects, and that is what accounts for their being characterized by the adjectives.

Third, the subject is given a kind of priority over the individual attribute. Primarily, this priority is due to the fact that adjectives demand subjects, but it is also mirrored in an incidental feature of the theory attributed to Protagoras, namely that the subjects are the progenitors of attributes.

Evidently the first two doctrines parallel Aristotle's theory of *paronymy* ('Brave things are so-called after bravery' [*Cat* 1a14-15]) and his theory of inherence ('A body is said to be white because the white is in it' [*Cat* 2a31-2]). The main difference between Plato's exposition in the *Theaetetus* and Aristotle's in the *Categories* is that Plato is concerned here with a thing's coming-to-be white, whereas Aristotle is concerned with its *being* white. This is due to an incidental feature of the theory that Plato is expounding, namely that Protagorean relativism supposes the world to be in flux. The third doctrine parallels Aristotle's doctrine of the *priority of substance*.

Many of the leading features of the *Theaetetus* account recur in the famous passage of the *Timaeus* (49-52) where the Receptacle is introduced in order to provide a framework for change. The central idea of this passage may be summarized as follows. In any change something fails to persist. But the subject itself persists through the change: to say that X changes implies that X persists. Therefore what persists must be an attribute.

Plato applies this idea to a problem concerning 'fire and the others.' Fire condenses to water, which evaporates to become air, which in turn becomes fire through combustion. Each element fails to persist through these changes — when fire condenses to water there is no fire left behind. So each element should be regarded as an attribute of some underlying subject, as a such (τοιοῦτον) not a this (τι) (49d). The underlying subject of these changes is the Receptacle (51a). The characters that inhere in this Receptacle, fire, etc., are transient copies of eternal forms — when some part of the Receptacle receives such a copy, that part is made fiery or whatever. Thus we have the theories of inherence and paronymy repeated from the *Theaetetus* — a copy of fire enters a part of the Receptacle and makes it fiery.

The above analysis of predication distinguishes two things — an individual attribute that comes and goes, and a subject that receives it, a 'this.' As we suggested earlier, the question 'What is the this?', or better 'Which is the this?', is a request to distinguish, in a manifold of individual attributes, between the things that inhere from those that constitute the Receptacle in which the others inhere. As we have seen, Plato sets up the problem in such a way that only a quality that persisted forever could be part of the Receptacle — for any alteration with respect to a quality puts it into the class of things that inhere. But *no* quality persists forever; even fire and water change into one another leaving no qualitative residue. So no quality is part of the Receptacle. The Receptacle is completely characterless. However, the form-copies do not occupy the whole of the Receptacle, but only parts — thus it is possible for one part of the Receptacle to be fiery and another simultaneously to be airy. So the form-copies may be said to occupy, and thus to characterize, *places*. These places are presumably not quite characterless — they have dimensions. So there is a kind of subject that is not intrinsically featureless. Plato believes that the dimensions that bound a place constitute subjects. Places have dimensions intrinsically, and come extrinsically to be characterized as *F* if occupied by individual instances of *F*-ness.[25]

On my interpretation, it will be noted, it makes perfect sense for Plato to say on the one hand that the Receptacle is characterless, and on the other hand that it is fiery. For what he means by this is that it has no characters intrinsic to it, while it is fiery only by paronymy. This marks an advantage over other interpretations which make this a contradiction. It will be noted that this advantage can be extended

---

25 At *Metaph* 1028b16-18 Aristotle refers to the view that 'the limits of body such as plane and line and point and unit are substances, more so than body and solid.' This is a good description of Plato's *Timaeus* view on my reading of it (with the admixture of Aristotelian terminology so typical of Aristotle's account of his predecessors). Interestingly enough, Aristotle identifies the view described at 1028b16-18 as Platonic (without explaining the reference).

to other discussions in the philosophical literature: for example, Descartes' discussion of the wax in the second *Meditation*, and Locke's notion of a substratum. The idea that a thing has its intrinsic properties as parts of itself, and is thus immutable with respect to these, is a powerful one and is to be found in many historical authors.

V Aristotle on Subjects

Plato's theory that places are subjects is open to a devastating criticism: it is unable to allow that some changes are locomotions and other alterations. Suppose that I move from p_1 to p_2. In doing so I take my individual attributes with me (all, that is, except my 'where'): I have changed my place but not my individual attributes. But Plato cannot accommodate this simple intuition. According to him, p_1 was initially filled with individual attributes and is now empty of them – so p_1 has *altered*. Similarly, p_2 has become filled with individual attributes – it too has changed. Furthermore, there is nothing, according to Plato, that has moved from p_1 to p_2 – for places are the only things. Thus there is no locomotion.

In order to evade this difficulty, Aristotle had to make places entities that retain their identity in spite of being occupied and vacated by substances – in other words he had to make a place not a subject but that *in which* a subject is. In other words, Plato's places must now be filled with subjects instead of being the subjects. But now recall Plato's thesis that if a subject changes with respect to a property F then F does not characterize it intrinsically. Now that Aristotelian subjects are capable of movement, there are no properties at all to characterize them, not even location. If Plato is right, and if subjects are eternal, then subjects would have to be *totally* characterless.

Characterless subjects do have a role in Aristotle's metaphysical scheme – prime matter is characterless – but they cannot be primary beings, he says, for featureless entities are not self-subsistent (*Metaph* 1029a27-8). A substance *is* a self-subsistent subject and thus it must have some intrinsic character. Still, it is conceded that, in the sub-lunary sphere at any rate, any substance may lose any property. Aristotle accommodates this fact by holding that when I belongs to a substance intrinsically, and the substance loses I, then the substance ceases to exist. He modifies the scheme of the *Timaeus* by allowing that subjects may perish.

In implementing this change, Aristotle reverses a decision Plato made in the *Timaeus*. A major difference between the *Theaetetus* framework for analysing subject-predicate statements and the one we find in the *Timaeus* is that in the latter we find a division of *things* themselves into subjects and attributes, whereas in the *Theaetetus* the distinction was relative to a particular perceptual encounter. There is nothing in the *Theaetetus* account which prevents *fire*, say, being a parent in one perceptual encounter, and so a subject for some attribute, and being an off-spring of another encounter, thus imparting a fiery quality to something else by its own birth. But this is precisely what is not allowed in the *Timaeus*: the question asked there is whether fire is a this or a such, and it is found to be a such. Period.

It is the *Theaetetus* possibility of a subject for one thing being itself an attribute for another, or rather a variation of this framework, that Aristotle exploits in the central books of the *Metaphysics*. According to Aristotle, a self-subsistent substance has predicative structure — its existence consists in some form being predicated of underlying matter.[26] If this matter is itself intrinsically featured then it is again analysed in the same way. Thus consider a human being: he or she consists of form predicated of a body organized in a certain way. The body itself consists of form predicated of an underlying substratum, in this case the homoiomerous organic substances — bone, flesh, and so on. So consider the form of the body: it is predicated of something, and thus becomes a component of a subject for further predicates.

This chain of analysis terminates in prime matter (*Metaph* 1029a20-6). But prime matter is not a substance, because it is not self-subsistent. At some point in the chain, then, we will find form being predicated of something that is not a substance.

As I said earlier, Aristotle rejects the Platonic thesis that every subject must survive every change. But he continues to exploit the idea that every change involves an unchanging subject. Again, he rejects the idea that subjects are virtually featureless, but he continues to exploit the idea that everything else is built up from a featureless substratum by an accretion of attributes. It is because of the latter idea that Aristotle's individual substances differ from mere bundles of attributes. They are bundles with predicative structure — distinguished into layers of subjects-with-attributes that constitute new subjects.

---

26 For discussions of the notion that form is predicated of matter see Joan Kung, 'Can Substance Be Predicated of Matter?', *Archiv für die Geschichte der Philosophie* **60** (1978) 140-59; and Carl Page, 'Predicating Forms of Matter in Aristotle's *Metaphysics*,' *Review of Metaphysics* **39** (1985) 57-82.

Prime matter receives attributes. After the addition of several such attributes we get a new substance. Further new attributes are predicated of the new substance, not of prime matter. (In fact the old attributes are too, as we shall see in the next section.) After a time we get a new substance, and so on. A substance like a man is composed of many such layers. (See Furth's intervention in the discussion of Marc Cohen's paper above, p.124, for a diagram that illustrates this idea.)

To say more about the structure of substances requires that I should say more about Aristotelian predication, and I shall do this in a moment. But first let me address a point recently made by Alan Code. Code writes:

> [The substantial form] is not predicable of a plurality of separable particulars. Rather it is predicable of matter, and matter is not a "this" ... It is clearly *not* to be identified with a *Categories* universal. (18)

The trick in answering Code is to distinguish between the relationship that a form has to matter, and the relationship that form has to the form-matter complex. What seems to me true is that forms need not be *in* any this — they are in matter, and matter is (at least sometimes) not a this. However, universal forms are, according to my account, *said-of* many thises. For the relationship between form and matter is just that between an accidental (κατὰ συμβεβηκός) attribute and its subject. But in coming-to-inhere in matter, universal forms spawn individuals. There is a predicative relationship that holds between these new individuals and the forms, for the individuals bear both the name and the definition of the form. Thus my account yields, in this respect, exactly what the *Categories* would ask for: secondary substances are universal because they are said-of a plurality of individual substances.

Further, because a universal predicated of an individual results in an individual attribute, the form predicated of matter results in an individual form. It is these individual forms that Aristotle identifies with individual substances. Note that these are not individual forms in the sense of forms that vary *qualitatively* from individual to individual. They are individual only in the sense that they are distinct instances of a universal: they are only numerically distinct.

VI Hylomorphic complexes

The demise of Plato's *Timaeus* theory of predication has another immediate effect. Where Plato analysed predication in terms of an individual instance of a universal being located in the place that constitutes the subject, Aristotle cannot. Aristotle may concede that, in the case of sensible substances at least, the individual attribute is in fact located where the subject is, but for him the fundamental relation is between the subject and the attribute, not the place and the attribute. For Aristotle, the unity that holds a manifold of individual attributes together can no longer be understood spatially. Aristotle understands this unity as irreducibly predicative. What constitutes a manifold as one is, quite simply, that all its members characterize a single substance.

It is this notion that we need predicative relations to characterize the unity of a manifold that distinguishes Aristotle's theory from other bundle theories. Like other bundle theories, it holds that every individual is associated with a bundle of individual attributes — this bundle is what I have been calling the 'manifold.' But unlike many bundle theories — Plato's, for example, or Hume's — it does not hold that the manifold *is* the substance, nor does it hold that there is nothing more to an attribute's characterizing a subject than its being present in the manifold associated with that subject. The claim is, by contrast, that a manifold is unified by the predicative relations that hold amongst its members, and that these predicative relations are not reducible to mere presence, or, in other words, to mere spatiality. Thus the category distinctions that are associated with predication are maintained within the manifold.

The notion of a unity constituted by predicative relationships is the motivation for a class of Aristotelian entities that I have elsewhere entitled 'predicative complexes.' A *predicative complex* is, in the first instance, an entity that consists of an individual attribute and a substance, and it exists exactly when the substance is characterized by the attribute. Thus the predicative complex consisting of Socrates and an individual instance of paleness—this is called 'pale Socrates,'[27] as we shall

---

27 In *Metaph* E 2, Aristotle says that to investigate whether Coriscus and musical Coriscus are the same or different is the task of the philosopher. In my view he arrives at the position that they are different, though accidentally the same.

see in a moment — comes into existence when Socrates comes to be pale, and perishes when he ceases to be pale (*Metaph* 1037b12-18).[28]

Predicative complexes fall outside the category structure because they are not simple entities, and only 'things said without any combination' fall into one category or other (*Cat* 1a25-7). Predicative complexes are, according to Aristotle, the referents of phrases like 'white man' and 'white Socrates.' These entities account for the material equivalence, in Aristotle's philosophy, of sentences like 'Socrates comes-to-be (or is) white' and 'White Socrates comes-to-be (or is),' because they come-to-be when a universal comes-to-be true of a substance. For example, when universal *white* comes to be true of Socrates, the predicative complex *white Socrates* comes-to-be and, as a part of it, an individual white. (More about the individual white later.)

By an extension of the idea articulated in the above paragraph, we can envisage predicative complexes that consist of two items in categories other than substance, provided that these characterize the same substance. Thus the 'white musical' will be a predicative complex provided that there is some substance that is both white and musical. By a further extension, we can think of predicative complexes that consist of more than two items, provided that these characterize the same substance. In this way, the 'manifolds' of individual attributes that I have been appealing to are predicative complexes, since they are collections of individual attributes that characterize a single substance.

Now when paleness comes to be true of a body, we have, by the above schema, two new individuals — an individual instance of paleness, and the predicative complex, the pale-body. In addition, we have a thing that persists through the change, the body. But there are certain changes, substantial generations, which are a little more complicated. In such changes, substantial form comes to characterize matter. Here we have a new predicative complex alright, the form-matter complex, but the wrinkle is that this new complex *is a new substance*. Now, as in the previous case, the coming to be of the form-matter predicative complex is accompanied by the coming to be of an individual

---

28 For the notion of a predicative complex, see my 'Greek Ontology and the "Is" of Truth,' *Phronesis* **28** (1983) 113-35. I now want to retract the suggestion, made in that article, that individual attributes are identical with predicative complexes (130), for I now regard individual attributes as components of predicative complexes.

instance of the attribute. This latter entity is, I contend, substance in the sense of form. The former entity, the form-matter complex, is composite substance — the hylomorphic complex.

The Aristotelian individual substance is not what we take it to be today — a concrete object with color, temperature, and the rest. Rather, it is the predicative complex that corresponds to the formula of the substance, together with the matter of which that formula is predicated. In other words, it is not the entire manifold of properties that is associated with a substance. It includes only the intrinsic, and none of the extrinsic characteristics of a substance.

VII Individual Form and Matter

My account of individual substances, as it has so far been articulated, makes form and matter extrinsic to one another, and united only within the composite substance. But this is somewhat inaccurate, and in this last section I shall attempt to show in what way there is a tighter connection between the form and its matter. In fact, as we shall see, there is a sense in which Aristotelian matter is unique to the substance to which it belongs — it comes into being when the substance does, and what remains when a substance is destroyed is not the matter of that substance.

According to Aristotle, every non-substantial universal has a 'primary recipient' — a sort of thing in which it must, by definition, be present. For example, *surface* is the primary recipient of *white*: that is, white is never present except in a surface. Primary recipients are not necessarily substances; for example, surfaces are quantities (*Cat* 5b2-6). Where the primary recipient is not a substance, it will itself have a primary recipient, so that there will be a chain of primary recipients between a universal and its subject. Color is received in surface, surface in solid, solid in body.

Substantial forms too have primary recipients. Aristotle sometimes talks as if matter is the primary recipient of substantial form. One might take him to mean, for instance, that the bricks and mortar out of which a house is constructed is the primary recipient of the form of a house. But that could not be quite right. For the form of house is *covering*. This form is often instantiated in bricks and mortar, but it could also be instantiated in stone, or in timber. All that is required is that it be instantiated in something rigid and durable. So *rigid and durable material* is the primary recipient of the form of *house*. At the same time, no other property of brick is intrinsic to the house. Thus it is the predicative

complex consisting of *covering* and *rigid and durable material* that constitutes the house, not the predicative complex consisting of *covering* and
bricks and mortar. A house, therefore, understood as a composite substance, is the form in the *proximate matter*: it is a covering made of rigid
and durable material.

Now this is an interesting result. Consider the process by which
the house came to be. To start with there is a pile of bricks, and some
mortar. Each of these things is a self-subsistent thing, a substance. Consider a brick. It has various intrinsic properties — whatever constitutes
its brickiness. As well, it is durable and rigid — it is not immediately
important whether this property is intrinsic or not, whether it is part
of the hylomorphic complex that constitutes the brick.

Now suppose that the bricks etc. are put together into a house. The
rigidity and the durability of the bricks are appropriated by the predicative complex that constitutes the house. That is to say the *house* will
now be durable and rigid. Further, durability and rigidity are *intrinsic*
properties of the house; so if they are intrinsic to the brick, their status remains unchanged. But this will not generally be true: not all the
intrinsic properties of the brick will be intrinsic properties of the house.
This much is implied by the bare observation that *bricks* are not required
for houses — stone will do. Thus it is only the proximate matter that
is intrinsic to the house. Still, most of the properties that characterized the bricks continue to characterize the house — it is rough, red,
etc. Even the intrinsic properties of the brick will continue to characterize the house. For example, if it is intrinsic to the bricks that they
contain clay as a constituent, that is a property of the house too — it
simply is not an intrinsic property of the house. But what this means
is that the properties of the brick are redistributed. Some of its intrinsic properties are extrinsic to the house, some of its extrinsic properties might be intrinsic to the house — and of course, some of the bricks'
properties vanish altogether.

But this means that the brick *no longer exists* — the unity that constituted it is no longer a unity. (Of course it does exist *potentially*.) Thus
one has to qualify the claim that the bricks persist through the change
that results in a house coming-to-be. And we have already seen that
it is not the bricks that constitute the subject for the form of the house.

An individual substance is, on my interpretation, an individual instance of a substantial form together with its primary recipient. The
coming-to-be of a substance involves two pre-existing things — universal form and matter. (In the case or self-reproducing substances an individual instance of the form is also required as a carrier of the
universal.) But when the form comes to characterize the matter the

result is just the composite substance and the individual form. The uninformed matter does not exist any longer as a complete or self-subsistent entity, because it has now been broken up, and has lost its substantial unity. The part that is the primary recipient for the substantial form is now a part of the individual substance. The remainder now constitutes properties that inhere in the individual substance. Thus it is 'only homonymously' that the matter can be identified as present in an individual substance.

The above account also shows why the matter is not ontologically prior to individual substance. For if the preexistent matter were the very same thing of which substantial form were predicated, it would be causally and temporally prior to the individual substance — something that seems inconsistent with Aristotle's general scheme of things. But now we see that that which pre-exists is not the matter that is part of the composite, and that which is part of the composite is not prior.

****

I started by saying that I would sketch a theory of substance within which the assertion that substances are 'what-is-it and this' would make sense. I claim to have done this. I have tried to motivate the distinction between intrinsic and extrinsic predication that lies behind the question 'What is substance?'. I have talked about the origins of the idea of substance as subject, and shown why Aristotle rejected Plato's idea that to be a subject was what it was to be a this. I have tried to explain why substances are thought of as entities with an internal structure, and why the relation between the elements of this structure is thought to be predicative in nature. And I have tried to show how substances are a special sort of predicative complex. Along the way, I have demonstrated several reasons for thinking both that substances are individual, and that they are what-is-its.

Analytica Posteriora

Categoriae

de Anima

de Caelo

Meteorologica

> 351b8: Supports the non-historicity of science, 29
> 390a7-19: *see de An* 412b14-15

Physica

> 194b1-7: Skills governed by purpose, which are therefore the source of ar-
> tefacts, 26
> 195a32-5, and *de An* 418a21-3: That substances are said to supervene on
> causally relevant non-substances shows directness of causal links in-
> volving non-substances, 161
> 195b6-20: Illustrates that problem of inertia had not been formulated, 31n
> 195b17-21: That τὰ ἐνεργοῦντα perish simultaneously with their effects shows
> that they must be individual attributes, 160
> 198a25-b9: The point is that formal cause functions as final, not that final
> furnishes formal, 27n
> 198b5: Seldom noticed summary of interrelation of the four causes, 27n
> 199a20-31: Illustrates that purposiveness imitates natural regularity, not
> vice versa, 28n
> III 1-3: Analytic framework for change requires there to be individuals in
> non-substantial categories, 160
> 201a10: Definition of change illustrates the non-identifiability of happen-
> ings until they are over, 30n
> VIII 1: *see de Gen et Corr* 338a15

Politica

> 1256b20-2: Interpolation by over-enthusiastic Judaeo-Christian commen-
> tator? 38n
> 1260a12-17: *see Metaph* X 9

Meteorologica

de Partibus Animalium

Physica

Politica

Topica

Aquinas